Dialogue, Science and Academic Writing

Dialogue Studies (DS)

*Dialogue Studie*s takes the notion of dialogicity as central; it encompasses every type of language use, workaday, institutional and literary. By covering the whole range of language use, the growing field of dialogue studies comes close to pragmatics and studies in discourse or conversation. The concept of dialogicity, however, provides a clear methodological profile. The series aims to cross disciplinary boundaries and considers a genuinely inter-disciplinary approach necessary for addressing the complex phenomenon of dialogic language use. This peer reviewed series will include monographs, thematic collections of articles, and textbooks in the relevant areas.

For an overview of all books published in this series, please see
http://benjamins.com/catalog/ds

Volume 13

Dialogue, Science and Academic Writing
by Zohar Livnat

Dialogue, Science and Academic Writing

Zohar Livnat

Bar-Ilan University

John Benjamins Publishing Company

Amsterdam / Philadelphia

The paper used in this publication meets the minimum requirements of American National Standard for Information Sciences – Permanence of Paper for Printed Library Materials, ANSI z39.48-1984.

Library of Congress Cataloging-in-Publication Data

Livnat, Zohar.
  Dialogue, science and academic writing / Zohar Livnat.
      p.   cm. (Dialogue Studies, ISSN 1875-1792 ; v. 13)
Includes bibliographical references and index.
1. Dialogue analysis. 2. Technical writing. 3. Academic writing. 4. Rhetoric. I. Title.
P40.5.D53L58   2012
401'.41--dc23   2011040343
ISBN 978 90 272 1030 2   (Hb ; alk. paper)
ISBN 978 90 272 7502 8   (Eb)

John Benjamins Publishing Co. · P.O. Box 36224 · 1020 ME Amsterdam · The Netherlands
John Benjamins North America · P.O. Box 27519 · Philadelphia PA 19118-0519 · USA

Table of contents

CHAPTER 1

Introduction   1

1.1   Purpose and structure of the study   2

1.2   Corpus and methodology   3

CHAPTER 2

Approaches to dialogicity   7

2.1   Dialogism and intertextuality: A discursive-literary approach   9

2.2   Language as dialogue: A communicative approach   15

2.3   Voices in the text: A linguistic approach   16

2.4   The speaker and his audience: An argumentative approach   18

2.5   Conclusions   19

CHAPTER 3

Academic discourse as persuasion   21

3.1   The persuasive goal of the paper's structure   24

3.2   The persuasive goals of research articles   28

3.3   Degrees of facticity   34

3.4   Argumentation, facticity, time: Three parallel lines   44

CHAPTER 4

The dialogic dimension of academic discourse   47

4.1   Towards a new model of scientific dialogicity   47

4.2   Citations   52

    4.2.1   Patterns of citations   53

    4.2.2   Authenticity and responsibility   59

    4.2.3   The rhetoric of citations   64

4.3   Concession   66

    4.3.1   Introduction   66

    4.3.2   The rhetoric of concession   73

    4.3.3   Concession as dialogue   88

4.4   Inclusive *we*   93

4.4.1   First-person pronouns   93
4.4.2   First-person plural in Hebrew   96
4.4.3   Inclusive *we* as dialogue   101
4.5   Questions   110
4.5.1   Direct and indirect questions   110
4.5.2   The rhetoric of questions   114
4.6   Scientific dialogicity: A combined model   120

CHAPTER 5
Scientific dialogicity in action   123
5.1   Introduction   123
5.2   The United Monarchy: A question from the past   127
5.3   The classic pattern   132
5.4   The conflicting pattern: Targeting a school   140
5.5   The conflicting pattern: Targeting a researcher   150
5.6   The ping-pong pattern   160
5.7   Face-to-face interaction   182

CHAPTER 6
Conclusions   195

Bibliography   199
Appendix: Corpus of journal articles   209

Authors index   213
Subject index   215

CHAPTER 1

Introduction

"If I have seen further it is only by standing on the shoulders of giants," wrote Sir Isaac Newton in a letter to Robert Hooke in 1676. This statement reminds us that the scientist, the researcher, the academic, even those working under conditions of complete isolation, whether in the laboratory or at one's desk, never operate in a vacuum. Relating to people and texts from the past and present is inevitable and largely shape the nature of their work. From the perspective of the history of scientific concepts, Fleck (1979) argues that "whether we like it or not, we can never sever our links with the past, complete with all its errors. It survives in accepted concepts in the presentation of problems, in the syllabus of formal education, in everyday life, as well as in language and institutions" (p. 20).

What was true for Newton is even truer for the researchers of today, both in the natural sciences as well as in the humanities and social sciences. The academic scholar, even if he feels that in certain respects he is a "one-man show," is part of a shared and ongoing endeavor, and his scientific work is a social act that exists within a framework of agreed-upon social mechanisms.

The scientific text or what is often called *academic prose* is often presented on the monologic-dialogic continuum as a classic example of a monologic text. Some linguists, in analyzing and comparing language variations, take it as being at the extreme monological pole, with spoken interaction being the other, most "dialogic" one (Chafe 1982; Biber et al. 1999).

However, a more in-depth scrutiny of the nature of scientific writing will show that it is only partially monologic in character. In fact, the texts that scientists write contain many dialogic features: They address other people in the past, present and future, relate to them and correspond with them in different ways. Moreover, it may be argued that scientific creativity, with the fluid and open-ended process that characterizes it, is rooted in an ongoing scientific conversation (Beller 1999: 2). Beller's analysis of the history of the quantum revolution in 20th-century physics is based on the notion of "dialogical creativity." According to her approach, "dialogical creativity is not an instantaneous "eureka" experience; it is rather a patiently sustained process of responsiveness and addressivity to the ideas of others, both actual and imagined" (ibid.: 6).

1.1 Purpose and structure of the study

The aim of this book is to investigate the dialogic nature of research articles from the perspective of linguistics and discourse analysis. In recent years, discourse analysts have examined various aspects of academic dialogicity, but no complete framework has yet been provided for these observations based on theories of dialogicity. That is why I undertook to propose in this book a theoretical and applied framework for the understanding and exploration of academic dialogicity, in which varying levels and types of dialogicity are expressed. This task will involve a number of different steps, which will be included in this book's two theoretical chapters (2 and 3) and two analytical ones (4 and 5).

Many current theories touch in one way or another on the question of the dialogicity of discourse from various perspectives, such as the cognitive, historical and communicative aspects. For the design of the theoretical framework of this study, I have chosen four theoretical approaches that appear to make the most substantial contribution to the research question that interests me here – the dialogicity of scientific texts. In Chapter 2, I will present each of them and then integrate them so as to best serve the analysis in the analytical chapters.

My assumption is that persuasion is by definition dialogic, since it requires orientation toward an audience in order to increase its acceptance. Thus, the next step will be to define the discourse used in research articles as a persuasive genre. In Chapter 3, I will support this definition in various ways through the analysis of excerpts taken from articles published in social sciences journals.

Chapter 4 will be devoted to the search for a coherent model of dialogicity that can serve as a solid basis for the analysis of scientific dialogicity. Four textual components that have been shown to be of considerable importance in designing the dialogic nature of academic discourse will be discussed: citations, concession, the inclusive *we* and questions. Further to this analysis, Section 4.6 will propose a combined model of scientific dialogicity that describes the place and role of these linguistic structures against the background of the various theoretical approaches to dialogicity.

Taking this combined model as a basis, Chapter 5 will demonstrate how scientific dialogicity is realized in an actual academic dispute. In order to take a closer and more in-depth look at the dialogic dimension of the academic discourse, I will propose an analysis of a number of complete articles. The articles all focus on the same topic and hold a dialogue with one another, and consequently, provide a better understanding of the specific content to enable a more precise interpretation of each of the linguistic components that can be identified in the texts. My analysis will show how a scientific project is constructed step by step by means of a dialogue with its readers and discourse community. In this chapter, I will offer a number of different patterns of scientific dialogicity characterized by the different levels of the polemic held with the research world and other specific researchers – from the "classic," moderate and polite dialogicity to a direct and personal confrontation between scientists.

1.2 Corpus and methodology

This study takes a two-fold course: It starts by exploring and describing scientific dialogicity in general, and continues by defining different types of dialogicity according to various measures of directness and confrontation. It is therefore based on two different kinds of corpora: One is a representative, controlled corpus of Hebrew research articles in the social sciences, and the other is a corpus of English-language articles in the humanities chosen in advance.

Chapter 4, which presents some of the main linguistic elements that construct the academic discourse as dialogue, is based on a corpus made up of 30 articles reporting quantitative and qualitative studies from various fields of social sciences, published in Hebrew in a refereed scientific journal. Israeli researchers in the social sciences are required to publish in English and are evaluated in accordance with the international recognition they earn, which is made possible only through the reading and writing of English (see Hamel 2007 for a critical view of this situation).[1] For that reason, Israel does not have a wide selection of Hebrew-language refereed periodicals in these fields. For the present study, I have chosen *Megamot*, the most prestigious Hebrew-language journal and the one rated as the best platform for scientific publications in the social sciences in Israel. It is a peer-reviewed journal that maintains the highest standards of scientific writing, which publishes papers on various subjects, such as education, sociology, anthropology, psychology, political science, administration, organizational behavior, communication and more.

My work is not comparative in nature in the sense that I do not have comparative corpora of corresponding Hebrew and English texts for example. However, in those instances where possible, I have presented and explained my findings in the context of other researchers' findings from the English.

---

1. The only area of Israeli academia that differs in this respect are those disciplines that can be included under the heading of "Jewish studies," in which Modern Hebrew is a lingua franca alongside English, whose importance varies in accordance with the fields of study and the modes of scientific communication. However, in the past few decades, English has been growing increasingly dominant even in these fields, at the expense of Hebrew (Sasaki 2007).

As may be expected, a large proportion of the discourse analysis studies dealing with scientific discourse analyzed English-language texts; however, a great deal of knowledge is now available regarding the features of this genre in other languages too. Studies available in English have described the features of scientific writing in French (Salager-Meyer 2001; Salager-Meyer et al. 2003; Fløttum et al. 2006; Tutin 2010), Spanish (Burgess 2002; Salager-Meyer et al. 2003; Martín-Martín 2005; Lafuente-Millán et al. 2010), Chinese (Taylor and Chen 1991; Bloch and Chi 1995), Finnish (Mauranen 1993), Norwegian (Fløttum et al. 2006), Swedish (Fredrickson and Swales 1994), German (Resinger 2010), Italian (Giannoni 2005), Polish (Duszak 1994, 1997), Czech (Čmejrková 1996, 2007), Russian (Čmejrková 2007), Malay (Ahmad 1997), to mention only a few. However, the Hebrew scientific or academic discourse has never been investigated from the point of view of discourse analysis, and to the best of my knowledge, my works are the first to describe its features.

In Chapter 5, which presents scientific dialogicity in action, the selected corpus is not a representative sample of any type, but rather a corpus that was selected in advance, deliberately to explore the manner in which the scientific project is constructed one step at a time by means of the dialogue with the discourse community. In this chapter, I have chosen to analyze papers originally written in English, in order to enable the reader to see entire sections of texts, with all the attendant linguistic nuances, without requiring the mediation of translation. Unlike studies that have investigated academic discourse by means of computerized corpora, or which restricted themselves in advance to seeking out specific linguistic features, I have decided to use the methodology of open and free discourse analysis in this chapter, to observe the entire text in its context, as well as to identify an entire range of linguistic and rhetorical strategies that aid the author in achieving his aims. In order to contribute to an understanding of the context, I have selected papers all of which deal with more or less the same issue, so that readers can advance their understanding of the matter at hand as they read. This is also why my analysis follows the outline of the analyzed papers themselves, so that readers can gain at least a partial understanding of the range of arguments offered in the paper.

All the papers analyzed in Chapter 5 come from the subject of archeology and ancient history, an interesting field from the perspective of its research methods. While defined as being part of the humanities, it is not a "textual" field of research, like philosophy or literary research. Archeology is a science that requires arduous, expensive field work involving many years and many participants, which ultimately leads to findings that may be viewed as "hard" data. However, these findings do not generally "speak for themselves": they require interpretation, and synthesis between them and historical knowledge from various other sources.

These sources are ancient texts that call for philological, textual and historical interpretation. The scientific claims in this field are then based on a combination of methods that differ in nature, each of which makes its own unique contribution to the research. In the analysis of the texts, we will see how the researchers themselves present the weight that they choose to assign to the various methods and their own preferences in regard to research methods.

CHAPTER 2

Approaches to dialogicity

It is possible to argue that the nature of written language as such is more mono-logic than dialogic. One of the basic differences offered by Chafe (1985) between written and spoken discourse is that while speaking typically takes place in an environment of social interaction, writing is a lonely activity. Writers are usually isolated from their audiences, both spatially and temporally (198:105, 116). This does not mean that dialogue and interaction mean only turn taking in conversa-tion. Nystrand (1986) argues that "turn taking is not interaction *per se* but merely the way conversants accomplish interaction" (p. 40). In fact, interaction takes place every time the reader understands a written text. Thompson and Thetela (1995) formulate the continuum between monologic and dialogic forms of dis-course in terms of a "continuum from most to least explicit forms of realization: that is, the speaker/writer may appear in the text […] with greater or lesser degree of visibility" (p. 109). Nevertheless, the dialogue between them is admitted by Thompson and Thetela to be at the essence of any language use.

The basis for including the interactive level in the description of meaning and function of any text is the belief that language use is only an example of the dialogi-cal actions that are essential for human life. Taylor (1991), in his inspiring essay "The dialogical self," explains the significance of dialogicity for any human being:

> We cannot understand human life merely in terms of individual subjects, who frame representations about and respond to others, because a great deal of hu-man action happens only insofar as the agent understands and constitutes him-self or herself as integrally part of a 'we.'
>
> Much of our understanding of self, society and world is carried in practices that consist of dialogical action. I would like to argue, in fact, that language itself serves to set up spaces of common action, on a number of levels, intimate and public. This means that our identity is never simply defined in terms of our in-dividual properties. It also places us in some social space. We define ourselves partly in terms of what we come to accept as our appropriate place within dialogi-cal actions. (Taylor 1991:311)

The model suggested by Sweetser (1990) for a global description of discourse structure assigns a major role to the interpersonal communication and the contact between the participants. Sweetser argues for a "multi-leveled cultural understanding of language and thought":

> In particular, we model our understanding of logic and thought processes on our understanding of the social and physical world; and simultaneously, we model linguistic expression itself not only (a) as description (a model of the world), but also (b) as action (an act in the world being described), and even (c) as an epistemic or logical entity (a premise or a conclusion in our world of reasoning).
> (1990: 21)

Thus, according to Sweetser, any discourse takes place in three domain simultaneously:

1. Content domain – the relations between entities and events in the physical world
2. Epistemic domain – the cognitive actions of the speaker
3. Speech-act domain – the utterance as a social act that appeals to other speakers

Although the three domains exist in all texts, the nature and degree of their presence vary across genres. In scientific discourse, the first domain is especially salient, whereas the last domain is relatively inconspicuous. The content domain is manifest in that the scientific discourse is mainly comprised of facts, knowledge and ideas, and the logical relations between them. The linguistic element mostly relevant to this domain is the excessive use of nouns and nominalizations, what Halliday (2004: 33) calls 'the nominal style.' According to Halliday, the wide distribution of nominalizations in scientific discourse has various causes or 'pay-offs.' First, they serve to 'construe technicality.' Since the nominal group is the most powerful resource for creating taxonomies (categories and subcategories), nominalization helps to create technical meaning through 'Classifier + Thing' structures (ibid.: 38–39). Second, by reconstructing qualities, processes and logical relations as 'things,' the grammar creates a semiotic universe of 'things' (ibid.: 47). Once qualities, processes and logical relations have taken on the feature of 'entity,' the researcher can observe, measure and experiment with them. "It is holding the world still, giving it stability and permanence" (ibid.: 129). Moreover, nominalization serves the movement from Theme to Rheme, which is characteristic of scientific discourse as a chain of reasoning. As 'things,' "they have the power of entering as participants into the full range of participant roles that the grammar has created for 'things'" (ibid.: 44). Nominalization "allows any observation, or series of observations, to be restated in summary form – compressed, as it were, and packaged by the grammar – so that it serves as the starting point for a further step in the reasoning" (ibid.: 19–20). The Theme is the stable part, and so it is typically construed as a noun, which is a result of packaging preceding information.

Nouns also have the power to present something as a fact through existential presupposition. The 'nominal' nature of scientific discourse is thus a significant phenomenon, and it is related, as I have explained, to the content domain of discourse. A further linguistic element that is relevant to this domain is the use of passive voice that emphasizes facts and ideas in the expanse of the speaker's presence (Livnat 2006).

The second domain – the cognitive actions of the speaker, is of particular interest in the scientific discourse. On the one hand, it is obvious that the scientific activity heavily relies on the intellectual effort of studying, planning, analyzing, interpreting, concluding etc. On the other hand, the scientific text does not give it a full expression, because it deliberately hides these cognitive actions through the use of impersonal expressions that diminish the presence of the author (Livnat 2006, 2010a). The epistemic domain thus is relatively implicit.

It is possible to argue, however, that the IMRD scheme (see Section 3.1 below) is a discourse structure that allegedly reflects the researcher's cognitive actions:

> She identifies a gap in knowledge, i.e. a niche that requires a filling → formulates a research question → plans how to answer it → interprets the findings → draws conclusions → closes the circle: answers the question and fills in the gap.

Although this description is a naïve, not to say erroneous description of the scientific work (as I argue below in Section 3.1), the IMRD scheme may be seen as a formal, schematic representation of these cognitive actions. From this perspective, the structure is an expression of the cognitive actions performed by the researcher.

The third domain – the interpersonal, dialogical and interactive domain – is the one that will be at the center of the present study that aims to describe the ways in which the dialogic aspect is manifest through various linguistic and discursive structures. In the present chapter, I will present a number of approaches to dialogicity that have their origin in different disciplines. Combining these approaches will comprise the theoretical framework and the ideological basis for discussing scientific discourse as dialogue.

2.1 Dialogism and intertextuality: A discursive-literary approach

The term "dialogism" is connected first and foremost with the Russian literary theorist Mikhail M. Bakhtin. In his literary theory and especially in his analysis of the discourse of the novel, Bakhtin extensively discussed the dialogic nature of language (1981:273). In his famous essay "Discourse in the novel" he writes:

> The transmission and assessment of the speech of others, the discourse of another, is one of the most widespread and fundamental topics of human speech. In all areas of life and ideological activity, our speech is filled to overflowing with other people's words, which are transmitted with highly varied degrees of accuracy and impartiality. The more intensive, differentiated and highly developed the social life of a speaking collective, the greater is the importance attaching, among other possible subjects of talk, to another's word, another's utterance, since another's word will be the subject of passionate communication, an object of interpretation, discussion, evaluation, rebuttal, support, further development and so on.
>
> (Bakhtin 1981:337)

The concept of dialogism goes far beyond the textual form that is usually called "dialogue." Bakhtin calls this form "the external, compositionally marked, dialogue," and distinguishes between this phenomenon and "the internal dialogization of discourse." According to Bakhtin, the phenomenon of internal dialogization is present to a greater or lesser extent in all realms of the life of the word (1981:284). Thus dialogism, for Bakhtin, is a constitutive element of all language (Allen 2000:21), and this is true even for what he calls "extra artistic prose" (everyday discourse, rhetorical discourse, and scholarly discourse), that "cannot fail to be oriented toward the 'already uttered,' the 'already known,' the 'common opinion' and so forth. The dialogic orientation of discourse is a phenomenon that is, of course, a property of *any* discourse" (Bakhtin 1981:279, italics in original).

Bakhtin discusses dialogicity from a number of interconnected aspects. One is that every speaker, by his very use of a certain word, gets involved in the dialogue surrounding this word that includes the ways other speakers used it before him. According to Bakhtin, the most important aspect of language is its reference to preceding utterances and to pre-existing patterns of meaning that anticipate future utterances. "Any utterance is a link in a very complexly organized chain of other utterances" (1986:69).

A specific reference to scientific discourse can be found in Bakhtin's work from time to time:

> However monological the utterance may be (for example, a scientific or philosophical treatise), however much it may concentrate on its own object, it cannot but be, in some measure, a response to what has already been said about the given topic, on the given issue […]. After all, our thought itself – philosophical, scientific, and artistic – is born and shaped in the process of interaction and struggle with others' thought, and this cannot but be reflected in the forms that verbally express our thought as well.
>
> (1986:92)

Our speech, according to Bakhtin, is filled with echoes and reverberations of other utterances. It is filled with "varying degrees of otherness and varying degrees of 'our-own-ness' [...]. Each utterance refutes, affirms, supplements, and relies on the others, presupposes them to be known, and somehow takes them into account" (1986:91). Words do not exist in a neutral and impersonal language. "It is not, after all, out of a dictionary that the speaker gets his words." Rather, they exist "in other people's mouths, in other people's contexts, serving other people's intentions: it is from there that one must take the word, and make it one's own" (1981:294). Not only the words, but also "the topic of the speaker's speech, regardless of what this topic may be, [...] has already been articulated, disputed, elucidated and evaluated in various ways. Various viewpoints, world views, and trends cross, converge and diverge in it" (1986:93), and there are always preceding utterances with which a given utterance enters into one kind of relation or another – builds on them, polemicizes with them, or simply presumes that they are already known to the listener (1986:69). Thus, "the subject itself inevitably becomes the arena where [the speaker's] opinions meet those of his partners (in a conversation or dispute about some everyday event) or other viewpoints, world views, trends, theories, and so forth (in the sphere of cultural communication)" (1986:94).

Bakhtin's discussion of dialogism is embedded in his discussion of genres. He stresses that all linguistic communication occurs in specific social situations and between specific classes and groups of language-users (Allen 2000:15). The stratification of language into various socio-ideological "languages" – languages of social groups, professional languages, language of generations and so forth – is a phenomenon that Bakhtin calls "heteroglossia" (1981:272). For Bakhtin, "all words have the 'taste' of a profession, a genre, a tendency, a party, a particular work, a particular person, a generation, an age group, the day and hour. Each word tastes of the context and contexts in which it has lived its socially charged life" (1981:293).

Another aspect of Bakhtin's "dialogism" comes out of his concepts of *addressivity* and *responsive understanding* that are related to the claim that in every kind of discourse, even a written text of a distanced and uninvolved nature, a dialogue with an addressee exists, and thus the discourse is shaped in accordance with the way the addresser perceives the addressee.

> An essential (constitutive) marker of the utterance is its quality of being directed to someone, its *addressivity* [...]. Both the composition and, particularly, the style of the utterance depend on those to whom the utterance is addressed, and the force of their effect on the utterance. Each speech genre in each area of speech communication has its own typical conception of the addressee, and this defines it as a genre.
> (1986:95, italics in original)

When the listener perceives and understands the meaning of speech, he simultaneously takes an active, responsive attitude towards it. "Any understanding of live speech, a live utterance, is inherently responsive, although the degree of this activity varies extremely." The speaker "does not expect passive understanding that, so to speak, only duplicates his own idea in someone else's mind. Rather, he expects response, agreement, sympathy, objection, execution and so forth" (1986:68–69).

When constructing an utterance, the speaker tries to act in accordance with the response he anticipates, so this anticipated response, in turn, exerts an active influence on his utterance. "The entire utterance is constructed, as it were, in anticipation of encountering this response" (1986:95). Thus, every single piece of the complexly structured works of secondary genres, such as a scientific text, is an 'utterance,' a unit of speech communication. It is "oriented toward the response of the other (others), toward his active responsive understanding, which can assume various forms: educational influence on the readers, persuasion of them, critical responses, influence on followers and successors and so on. [...] The work is a link in the chain of speech communication. [...] it is related to other work-utterances: both those to which it responds and those that respond to it" (1986:76).

Bakhtin's concept of *dialogism* is extraordinarily influential within various branches of modern thinking, including literary theory and criticism, linguistics, political and social theory, philosophy and many other disciplines (Allen 2000:15).

The term *intertextuality* first occurs in Julia Kristeva's early work of the middle to late 1960s, while she introduced the work of Bakhtin to the French-speaking world, and further developed the concept of dialogism while she placed a psychological dimension onto it. Despite not being a transparent term, it became one of the central ideas in contemporary literary theory and postmodern thought, exerting a considerable influence on other fields too.

Bakhtin (1986:72) hinted at two different dimensions of intertextuality that can be indentified in the text, the *horizontal* dimension and the *vertical* dimension, but he did not make a clear distinction between them. According to Kristeva, in the horizontal dimension "the word in the text belongs to both writing subject and addressee." In the vertical dimension, "the word in the text is oriented toward an anterior or synchronic literary corpus" (Kristeva 1980:66). Allen (2000) explains:

> Authors communicate to readers at the same moment as their words or texts communicate the existence of past texts within them. This recognition, that the horizontal and vertical axis of the text coincide *within* the work's textual space, leads one to a major redescription of Bakhtin's theory of the dialogic text which culminates in the new term, intertextuality.   (Allen 2000:39)

Horizontal intertextual relations are thus the relationship between a text and those which precede and follow it in the chain of texts, whereas vertical intertextual relations are the relationship "between a text and other texts which constitute its more or less immediate or distant contexts: texts it is historically linked with in various time-scales and along various parameters, including texts which are more or less contemporary with it" (Fairclough 1992: 103).

Kristeva, like Bakhtin, has worked mainly in the field of literary theory. Her work on intertextuality focuses heavily on late nineteenth-century and early twentieth-century avant-garde writing, when explicitly intertextual writing came to the fore. Kristeva argues that it is in Modernist texts that the intertextual, or "transpositional" (as she later called it) aspect, begins to be self-consciously exploited (Allen 2000: 55). However, intertextuality as a conceptual framework can be found in the works of discourse analysts that are interested in the ways various kinds of texts or genres are shaped in relation to social, cultural and historical contexts. Fairclough's work on "Discourse and social change" (1992) is one example. According to Fairclough, the "inherent historicity of texts enables them to take on the major roles they have in contemporary society at the leading edge of social and cultural change [...]. The rapid transformation and restructuring of textual traditions and orders of discourse is a striking contemporary phenomenon, which suggests that intertextuality ought to be a major focus in discourse analysis" (1992: 102).

Following Bakhtin's intensive interest in genres, Fairclough refers to the difference between intertextual relations to other texts and intertextual relations to conventions. Using a distinction proposed by Authier-Revuz (1982) and Maingueneau (1987), he defines two kinds of intertextuality: *Manifest intertextuality* is the explicit presentation of other texts in the text under analysis, whereas *constitutive intertextuality* is "the configuration of discourse conventions that go into its production" (Fairclough 1992: 104). The concept of *constitutive intertextuality* thus emphasizes that the intertextual aspect of a text is related also to its relationships with discursive and stylistic conventions that are derived from certain genre norms.

Another important elaboration of the general notion of intertextuality, still in the framework of literary theory, was suggested by Genette (1997). For Genette, the broader term is *transtextuality*, which stands for everything that positions the text in a relationship – whether explicit or implicit – with other texts. He recognizes five types of transtextual relationships, and lists them more or less in the order of increasing abstraction, implication and comprehensiveness.

The first type is termed *intertextuality*, which, in a more restricted sense than that of Kristeva, is defined as a relationship of co-presence between two texts or among several texts, namely the actual presence of one text within another. In its most explicit and literal form, it is the traditional practice of *quoting* (with quotation marks, with or without specific references). In another less explicit and canonical form, it is the practice of *plagiarism*, which is an undeclared but still literal borrowing. Again, in still less explicit and less literal guise, it is the practice of *allusion*: that is, an enunciation whose full meaning presupposes the perception of a relationship between it and another text, to which it necessarily refers by some inflections that would otherwise remain unintelligible (pp. 1–2).

The second type is the generally less explicit and more distant relationship that binds the text properly speaking, taken within the totality of the literary work, to what can be its *paratext*: a title, a subtitle, intertitle; prefaces, postfaces, notices, forewords, etc.; marginal, infrapaginal terminal notes; epigraphs; illustrations; blurbs, book covers, dust jackets and many other kinds of secondary signals, whether allographic or autographic. These provide the text with a (variable) setting and sometimes commentary (p. 3).

The third type of textual transcendence, which he calls *metatextuality*, is the relationship most often labeled as "commentary." It unites a given text with another, to which it refers without necessarily citing it (without summoning it), in fact, sometimes even without naming it (p. 4).

The forth type is *hypertextuality* – any relationship uniting a text B (which he calls the *hypertext*) to an earlier text A (which is called the *hypotext*), upon which it is grafted in a manner that is not that of commentary. The general notion is of a text in the second degree – a text derived from another preexisting text. This derivation can be descriptive or intellectual, where a metatext "speaks" about a second text. It may be of yet another type, such as text B not speaking of text A at all but being unable to exist, as such, without A, from which it originates and which it consequently evokes more or less perceptibly without necessarily speaking of it or citing it (p. 5).

The fifth type, the most abstract and most implicit of all, is *architextuality*, the entire set of general or transcendent categories – types of discourse, modes of enunciation, literary genres – from which each singular text emerges (p. 1). It involves a relationship that is completely silent (p. 4).

These five types of transtextuality are not discrete and absolute categories without any reciprocal contact or overlapping.

In discussing dialogism and intertextuality of academic discourse, relevant phenomena would obviously be the presence of past texts in the form of citations, the ways in which the academic text is addressed to the reader and the ways the scientific arguments are worded based on the attempt to anticipate the reader's response. It would be equally interesting, however to refer to conventional language patterns, conventional terminology (surrounding which there may indeed be a dialogue) and genre conventions, because their use by the writer is an expression of his or her membership in the discourse community and a way of shaping his or her dialogue with it.

2.2 Language as dialogue: A communicative approach

Another approach that may contribute a great deal to the description of scientific dialogicity is Edda Weigand's concept of "Language as dialogue" (Weigand 2009). Weigand rejects the traditional distinction between monologue and dialogue for not being able to describe the nature of language as a form of communication. Her theory is based on two premises: Language is primarily used for communicative purposes, and communication is always performed dialogically. Thus, language use is inherently dialogic. When we speak, we always have a communicative purpose that determines our linguistic action (Weigand 2009: 23–24). Any monologic element of language is a partial aspect of the dialogic phenomenon (p. 34). There is no individual act without dialogic orientation (p. 65).

The concept of "Language as dialogue" was developed as a response to the speech-act theory that presents the speech act as the minimal unit of communication. Weigand's more developed model is a dialogic classification of speech acts, which is based on two functionally different types of action:

a. An *initiative* speech act which makes a pragmatic claim;
b. A *reactive* speech act, which is expected to fulfill this claim. This reaction may also be a mere mental one.

The initiative and the reactive action together produce the minimal autonomous communicative unit. The mutual independence of the acts is due to their purposes, which are dialogically oriented towards each other. Dialogicity is thus anchored in the individual acts themselves (p. 64–65).

Weigand's classification of the pragmatic claims is based on a distinction between claims to truth and claims to volition. Representative speech acts (= Assertions) are defined as claims to truth, and thus they are said to be directed at a reactive act of fulfilling the claim to truth, that is, a speech act of Acceptance (p. 45–46).

The rational basis which determines the internal interdependence of the initiative and the expected reactive action, which constitutes what Weigand calls the "Dialogic Principle Proper," "lies in the very characteristics of the initiative speech act itself, in its pragmatic claim which is to be negotiated in dialogue. It is, for instance, rationally defined by the characteristics of the Representative itself as a speech act which makes a claim to truth that a reaction is expected which will take up this claim, accept it or reject it, or clarify its conditions" (p. 274–275). We do not express a Representative only for the representative itself; we express it for a communication partner including ourselves. Consequently, the speech act following the Representative are most adequately described as reactive speech acts of ACCEPTING (p. 78). "In the same way, it is rationally defined by the characteristics of the Directive as a speech act which makes a claim to volition to which a reaction of Consent is expected. This internal interdependence of action and reaction is constitutive for every type of action and results in a *dialogic speech act taxonomy* which considers speech acts to be components of minimal dialogues" (p. 275). Thus, we arrive at the following global speech act types:

REPRESENTATIVE	↔	± ACCEPTANCE
DIRECTIVE	↔	± CONSENT
EXPLORATIVE	↔	± RESPONSE
DECLARATIVE	↔	[+ CONFIRMATION]

These pairs constitute the minimal communicative 'action games.'

The implication for academic discourse is that the arguments presented in the paper are transmitted not monologically, but rather dialogically directed at someone, with the expectation of a reaction of either acceptance or rejection. From this perspective, they are initiative speech acts that create expectations for a response from the reader: acceptance or rejection.

2.3   Voices in the text: A linguistic approach

From a linguistic perspective the question of 'voices' in the text has to do with the linguistic forms that are known as direct and indirect speech, reported speech and constructed dialogue, with which linguists have been dealing for a few decades now (see, for example: Wierzbicka (1974); the collection edited by Coulmas (1986); the special issue (no. 19) of *Text* from 1999). Thompson's model (1996) that I am presenting here is also a linguistic model. Its aims are more restricted than those of Bakhtin and Weigand, and it does not make an explicit claim regarding the nature of language as a whole. Thompson's purpose is to investigate the ways in which language events are reported in texts, and his work reflects specifically on English. Admitting a major inspiration of Bakhtin's concept of heteroglossia, he proposes a functional model that describes the various ways in which voices are signaled in text, that is, he is specifically interested in manifest intertextuality. What is defined by Thompson as "signaled voices in the text" is "any stretch of language where the speaker or writer signals in some way that another voice is entering the text, in however muffled or ambiguous a fashion" (Thompson 1996: 506). This category is much wider than what is usually included in "reported speech," but does not have the holistic nature of Bakhtin's or Kristeva's perception of the multiple-voiced language.

Thompson, like other discourse analysts and pragmatists,[2] assumes that there is a basic distinction between *averral* and *attribution* (Sinclair 1988). Averral is the default condition, where the addressee can assume that the responsibility for the proposition lies with the speaker. Attribution, on the other hand is the case where a proposition is indicated as deriving from a certain source. Broadly speaking, a text can be taken as averring anything that is not specifically attributed to another source. For Thompson, any case of attribution is marked, thus encouraging and justifying an investigation of the reasons why the speaker has chosen it.

Thompson's model proposes four dimensions in which the various voices in the text can be analyzed:

1. *The voice.* The voice that is presented as the source of the report can be of a specified or unspecified other(s); it can be a voice of a community; a voice of an unspecifiable other, where the question of the source is irrelevant because the formulation of the utterance deliberately leaves this question open in order to enable, for example empathy of the readers with the reported voice. It can also be the voice of the speaker himself from a different time and place.
2. *The message.* Utterances can vary in the degree in which they match the original utterance. The categories, from the closest and most similar to the most detached, are: quote, echo, paraphrase, summary and omission.
3. *The signal.* The ways in which the reporter can signal that a stretch of language is a report are varied. This dimension is mainly structural in nature and is based on the question of grammatical subordination. The signal may be separate from or fused with the reported voice. When it is separate from it, the signal may be grammatically dominant (reporting clause), subordinate (adjunct), or equal in status.
4. *The attitude.* The speaker can express his positive or negative attitude toward the truthfulness of the message cited or toward the source of the voice. The speaker's attitude may also be neutral and in some cases irrelevant in the context.

Looking at academic discourse from this perspective, it is relevant to mention at this stage some of the discursive realizations of these categories. Citations from

---

2. See, for example, Sperber and Wilson's (1981) distinction between 'use' and 'mention.' See also Goffman's (1981) notion of 'footing.'

research literature express the voice of a specific other, while the voice of a non-specific member of the discourse community can be heard through phrases such as 'some researchers' or 'in most books.' Self-quotations express the writer's own voice, and references to shared knowledge, conventions and norms of the disciplinary community express the voice of the community. These and other dialogic elements will be discussed in detail in Chapters 4 and 5.

2.4 The speaker and his audience: An argumentative approach

In the realm of argumentation or rhetoric, the concept of dialogicity has a unique meaning and significance. As Bakhtin put it, all rhetorical forms of language, even those that are monologic in their compositional structure, are oriented toward the listener and his answer. "This orientation toward the listener is usually considered the basic constitutive feature of rhetorical discourse" (Bakhtin 1981: 280). This is the reason for the major role of the audience in every argumentative or rhetorical framework.

In their "New Rhetoric," Perelman and Olbrechts-Tyteca (henceforth P&OT) defined the goal of all argumentation as "to create or increase the adherence of minds to the theses presented for their assent" (P&OT 1958: 45). By that very fact, any argumentation assumes the existence of an intellectual contact, namely, a contact between minds. Thus, the adherence of the audience is for P&OT a factor of psychological nature (p. 104), and the theory of argumentation, which, with the aid of discourse, aims at securing an efficient action on minds, might have been treated as a branch of psychology (p. 9).

The term "audience," which in classic rhetoric refers to the addressees of spoken speech, is extended in "The New Rhetoric" to include readers of a written text. "Whereas a speech is conceived in terms of the audience," write P&OT, "the physical absence of his readers can lead a writer to believe that he is alone in the world, though his text is always conditioned, whether consciously or unconsciously, by those persons he wishes to address" (1958: 7). For the purpose of persuasion, it is extremely important to form a concept of the anticipated audience, as close as possible to reality (p. 20). "It is indeed the audience which has the major role in determining the quality of argument and the behavior of orators" (p. 24).

The claim that the universal audience is a creation in the speaker's mind has been a constant matter in dispute and criticism (cf. Van Eemeren and Grootendorst 1995: 124). Jørgensen (2009) argues that the universal audience is not a subjective construction; rather, it is related to *intersubjective* rationality, explaining that "the main point in making the universal audience a projection made by the arguer, is that what counts as reasonable depends on time and place" (Jørgensen 2009: 12–13).

Besides their emphasis on the psychological aspect that is related to the mental connection with the addressee, P&OT refer also to the social aspect of argumentation:

> The study of audiences could also be a study for sociology, since a man's opinions depend not so much on his own character, as on his social environment, on the people he associates with and lives among. […] Every social circle or milieu is distinguishable in terms of its dominant opinions and unquestioned beliefs, of the premises that it takes for granted without hesitation.   (p. 20)

When the audience is part of a certain discipline, a crucial point is the disciplinary conventions and the knowledge shared by the members of the discipline:

> There are agreements that are peculiar to the members of a particular discipline, whether it be of a scientific or technical, juridical or theological nature. Such agreements constitute the body of a science or technique. They may be the result of certain conventions or of adherence to certain texts, and they characterize a certain audience. […] Initiation into a given discipline consists of communicating its rules, techniques, specific ideas, and presuppositions, as well as the method of criticizing its results in terms of the discipline's own requirements.
> (P&OT: 99–100)

The argumentation theory presented in "The New Rhetoric" is thus sensitive to the dialogue that the speaker holds both with a specific addressee as well as with disciplinary conventions. In fact, this approach assumes that persuasion inherently requires dialogicity. Specifically, it may be argued that the persuasive goals of academic discourse are achieved by drawing explicit and implicit dialogic relations with others: other researchers, other texts, the readers, the discourse community. Scientific arguments are always designed in accordance with what the members of the discipline know and believe, in compliance with the procedures and theories they accept, in accordance with the controversies that exist among them, etc.

2.5 Conclusions

Taking all these approaches together, we can now point to the main dialogic aspects and formulate them in such a way as to be helpful to us in analyzing academic discourse. We may say that discourse, any discourse, is dialogic from the following non-exclusive aspects:

1. It is addressee-oriented: It is designed according to the concept of the addressee that the speaker has, and aimed to anticipate his response.
2. It is responsive in the sense that it responds to former utterances and anticipates the next.
3. It is a mosaic of former utterances and linguistic usages.
4. It is a construction and reconstruction of various voices, specific and non-specific.
5. It is embedded in a specific genre, and thus expresses stylistic conventions, norms and shared knowledge of the discourse community.

This is the conceptual framework that serves me in the present study. On this background I will explore in Chapters 4 and 5 textual structures that are related to the presence of various 'voices' in the text: The author(s), former authors and researchers, readers, and the disciplinary community, its conventions, its norms and its shared accepted knowledge.

Taking the argumentative assumption that persuasion is necessarily dialogic as a point of departure, we may expect every argumentative text to reflect dialogicity of some kind. In order to justify a dialogic-argumentative analysis of academic discourse, we need to make an explicit effort to characterize this kind of discourse as an argumentative genre. I will engage in this task in the next chapter.

CHAPTER 3

Academic discourse as persuasion

Generally speaking, the scientific community is a community of people that share a set of common public goals (Swales 1990: 24), namely "the steady extension of the scope and precision of scientific knowledge" (Kuhn 1962: 52). The members of this discourse community also share norms, conventions, social structures and language. Nevertheless, this universe of discourse is far from being monolithic or uniform in form. The various disciplines in the academe are considered to be "subcultures" (Clark 1962) or "tribes" (Becher and Trowler 2001), each one having its own particular qualities, norms, practices and a relatively stable rhetorical situation (Hyland 1998: 20). The means by which arguments are presented, procedures enumerated, literature cited, theory and data discussed can only be seen as effectively persuasive against a backdrop of disciplinary practices and rhetorical expectations (Hyland ibid.). Research in academic writing in the past decade has established that academic discourse varies according to disciplinary conventions and cultural expectations (Hyland 2006). Research investigating English academic writing has demonstrated a long list of variations across disciplines, including argumentative moves (Holmes 1997; Lindeberg 2004), authorial stance (Kuo 1999; Bondi 2005; Groom 2005), speech acts (Myers 1992), pronouns (Kuo 1999; Harwood 2005; Fløttum et al. 2006), self-citation (Hyland 2001, 2003; Fløttum et al. 2006), hedging and mitigation (Hyland 1998; Vold 2006), critical and evaluative expressions (Lindeberg 2004; Stotesbury 2006), adversatives (Fløttum et al. 2006), engagement with audience (Swales et al. 1998), questions (Hyland 2002), negation (Fløttum et al. 2006), and meta-text (Samson 2004; Bondi 2005). Each discipline shares mechanisms of intercommunication among its members, such as professional journals and scientific conferences. The community's members have an in-depth familiarity with the types of texts that are unique to that community (Swales 1990: 25–26). The mastery of these genres is acquired over years of specialization and is intimately connected to the struggle waged by each new member to become a full-fledged member of the community.

The research paper is in many respects the most important genre, or at least the genre, mastery of which is most important to the professional success of the researchers. Their career is almost entirely dependent on their ability to consistently and continually write successful academic papers, publish them in the finest scientific journals and ultimately, cause others to cite them. Within its discourse community, this genre is a communicative tool that strives to attain social aims, and the form, structure and formulations it employs reflect the structure of the specific disciplinary community, its norms and conventions (Swales 1990).

This type of discourse has been defined as a form of persuasive writing and has been the subject of much research over the past decades by philosophers, linguists and sociologists of science, among others. It is agreed that the primary objective of the research article is not simply to present new claims, but to ensure that those claims are accepted and ratified as new knowledge by the disciplinary community (Hyland 1998: 25). The final rhetorical aim of a research article is to create effects that convince the audience to such a degree that the article becomes an integrated part of a particular field's literature (Fløttum et al. 2006). Prelli (1989) claims that "logic and experimentation are not the fundamental means of securing scientific change. They are efficacious only if applied *persuasively*" (Prelli 1989: 100; italics in original). Along the same lines, Berge (2003: 155) describes the scientific writer as "a rhetorician trying to establish the probability of his or her interpretation" (Berge 2003: 155).

Chaïm Perelman and Lucie Olbrechts-Tyteca (P&OT), in their treatise *The New Rhetoric*, argued that even scientific discourse requires argumentation, and described the establishment of the connection with the audience as the point of departure for this argumentation:

> The authors of scientific reports and similar papers often think that if they merely report certain experiments, mention certain facts or enunciate a certain number of truths, this is enough of itself to automatically arouse the interest of their hearers or readers. This attitude rests on the illusion, widespread in certain rationalistic and scientific circles, that facts speak for themselves and make such an indelible imprint on any human mind that the latter is forced to give its adherence regardless its inclination. (P&OT 1958: 17)

Pera (1994) in his analysis of scientific texts by Galileo, Darwin and modern cosmologists, uses the term "scientific rhetoric," arguing that rhetoric is central to the creation of scientific knowledge. He found that in scientific contexts, rhetorical arguments such as *ad hominem* and *ad ignorantiam* are used when the authors justify the choice a methodological procedure, interpret a methodological rule, apply a rule to a concrete case, justify their starting point, reinforce a hypothesis, criticize or discredit rival hypotheses or reject objections against a hypothesis (Pera 1994: 97–102). Despite the fact that scientists explicitly deny and reject the use of rhetoric, the analysis shows that they use typically rhetorical arguments in addition to deductive and inductive ones.

Looking at the scientific discourse as persuasion is further justified by the sociology of science. Sociologists, anthropologists and historians of science, such as Latour and Woolgar (1979), Knorr-Cetina (1981), Shapin (1984), Bazerman (1988), Beller (1999) have provided insightful descriptions of scientific texts and their role within the disciplinary discourse community from a social perspective. They sometimes have defined the writing of the scientific article as a social act that is part of a closed circle, one of whose roles is to enhance the researcher's position and reputation, so that he or she can persist in their research plan and obtain support for further research. Latour and Woolgar (1979) present scientists as having considerable skill in the art of persuasion. These skills:

> [E]nable them to convince others that what they do is important, that what they say is true, and that their proposals are worth funding. They are so skillful, indeed, that they manage to convince others not that they are being convinced but that they are simply following a consistent line of interpretation of available evidence. Others are persuaded that they are not persuaded, that no mediations intercede between what is said and the truth.   (pp. 69–70)

This is, according to Latour and Woolgar, the paradox of scientific writing: "The readers are only fully convinced when all sources of persuasion seem to have disappeared" (ibid.:76). They believe that the problem that writers of scientific papers deal with is to persuade the readers to accept their claims as a fact: "To this end rats had been bled and beheaded, frogs had been flayed, chemicals consumed, time spent, careers had been made or broken [...]" (ibid.:88).

Many linguists and discourse analysts have accepted this position and argued that the aim of academic papers is to persuade the scientific community to accept the new knowledge and the new arguments and make them part of the body of "scientific knowledge" or the "facts" regarding which there is a consensus within the relevant discipline. The text of the academic paper has been considered an argumentative text, whose primary objective is not simply to present new claims, but to ensure that those claims are accepted and ratified as new knowledge by the peers of the disciplinary community (Hyland 1998:25). The greatly increased competitiveness of modern science forces researchers to promote their own work, its significance and news value through their use of language (Huckin 1993; Lindeberg 2004).

The role of rhetoric in scientific writing is further clarified when we follow the various drafts scientists write until the final version of an article is submitted for publication. Bazerman (1988:Ch. 7) proposes an analysis of an article by the physicist Arthur Holly Compton from 1923, in which he proclaimed what is now known as the Compton effect, which is considered the first empirical verification of the quantum theory.

Bazerman examined all the early versions of the paper and the rewriting efforts by the researcher from the moment he began to write the article, and compared them to the final version as it was published in *Physical Review*. These texts reflect the author's dilemmas on all the relevant subjects: the choice of the broader context in which to present the problem, the choice of the order in which the matters would be presented, the addition of justification to a particular argument, the fine-tuning of definitions, the addition of details for the purpose of clarification or their removal when they appear to be superfluous, indecision regarding the addition of superlatives (in this case, praise for the equipment that he used).

Many differences between the versions are related to the author's presence in the text and the responsibility he demonstrates for his claims and for actions. For instance, the utterance "can leave no reasonable doubt," to indicate the researchers' evaluation is transformed into a more direct and personal formulation, "we believe establishes." A large set of revisions has to do with the degree of certainty the author has assigned to his claims: "Fact" is weakened to "observation," a "theory" is demoted to "hypothesis," and we find hedges and modals such as "about," "rather," "often" and "may be" (Bazerman 1988: 217). In the concluding section, where the most direct judgments are found, the various versions reflect the author's indecision that end up to a mitigation of his own judgments. In this way, according to Bazerman, the judgment is passed on to the audience (ibid.: 218).

It is also interesting to follow the changes that a paper undergoes in order to be accepted for publication. Myers (1985) examined different versions of articles on biology whose authors were forced to make changes following comments from the journal's reviewers. It would appear that the more definitive and unequivocal the wording of the arguments was, the more difficult it was to accept them and consequently, in the final versions, claims become more cautious, speculations and proposals more restricted and there is a considerable increase in hedging. From this perspective, the writing of a research article seems to involve "selective representation and rhetorical reconstruction as a means of anticipating negative responses to claims" (Hyland 1998: 18).

All the studies, attitudes and insights mentioned above present the scientific paper not as an expository, monologist, neutral and objective text, but rather as a persuasive, dialogic and polyphonic discourse that gives expression to a speaker and a reader as well as to the discourse community of which it is part.

3.1 The persuasive goal of the paper's structure

Do the text and the structure of the scientific paper reflect the way in which the research has actually been conducted? All the sociologists of science share the view that this is not the case. The "story" told in the paper may be even a reversed version of what in fact took place within the confines of the laboratory (Swales 1990: 118). In the lab, the scientists may opportunistically respond to an incidental finding, rather than consciously plan to try and solve a particular problem (Knorr-Cetina 1981).

A certain idea might come up as a circumstantial consequence of connecting two phenomena that share a weak link, and which was not considered significant in the first place. This connection may be used after other ideas fail. In the end, however, such an idea might be presented in the scientific paper as the product of logical reasoning, with the link between the phenomena transformed into a strong logical connection (Latour and Woolgar 1979: 173).

A major phenomenon that is responsible for the 'gaps' between the scientific work and the scientific text is the "conventional omission" (Dressen 2002). Huckin (2002) related to it in terms of "conventional silence," a silence that is governed by genre conventions. For instance, research reports routinely leave out methodological details. Dressen (2002) describes how most field results in geology disappear from the published research article during the act of "recontextualization" (Linell 1998). "The conventionalized silences of modern scientific discourse do not permit writers to reveal their story in explicit terms" (Dressen 2002: 89). Montgomery (1996) believes that the reader of the final product meets "a coherent, harmonious and successful investigation" (p. 13), while during the actual writing "guess-work and false starts, abandoned paths of logic, conversations and battles among participants and others, actual moments of insight, the pressure of deadlines, any and all emotional states or problems, and, of course, the detailed trials and tribulations of research itself" – are lost, erased or rendered irrelevant. The author is presented to us as an idealized writer that "provides a heroic objectivity" (Montgomery 1996: 18). Thus, "we find ourselves far away from a world in which it is expected that researchers will 'tell it as it happened,'" as nicely put by Swales (1990: 125). On the other hand, it is interesting to mention Beller's (1999) concept of 'polyphony,' which refers to the ways in which in a context of a scientific breakthrough, the published papers may reflect conflicting voices, disagreements, struggles and unresolved tensions among scientists working on the same issue.

Just as the language of the paper and the details selected for publication only partially reflect the "reality" of the research process, the same is assumed for the structure of the paper (Swales 1990: 118–121). Many academic papers are organized in the following four-part structure: Introduction-Method-Results-Discussion (= IMRD). It is a scheme that has been borrowed from experimental studies in Natural Sciences and applied to the "soft" sciences (Social Sciences and Humanities), and is used also for reporting on qualitative studies. However, this structure is not necessarily a real reflection of the thinking and working processes of the researcher, but rather a discourse pattern that has its own goals and functions. Hyland (1998) claims that this discourse pattern "falsely suggests that laboratory activity proceeds in a unidirectional, systematic fashion of experiment-result-interpretation-conclusion. Science rarely proceeds like that however" (Hyland 1998: 17). According to Gross (1985), this form, which moves "from the contingency of laboratory events to the necessity of natural processes," (ibid.: 20) is a rhetorical device that helps to justify the enterprise of experimental science:

> The introduction places the experiment in the context of a research program rather than laboratory events, thus creating a theoretical rather than contingent interpretive world, and this process is continued in Methods and Results where the objectivity of laboratory practices is emphasized over contingencies of human activities. Finally, the Discussion frames the relevance of these practices in the theoretical perspective of the Introduction, emphasizing a correspondence of the data to the paper's claim.   (Hyland 1998: 17, following Gross 1985)

I will cite one example from my corpus that helps to clarify some of these concepts. It is a statement justifying the researchers' choice of a particular research population.[3]

(1)   Since "job insecurity" in this study is general and does not refer to an organizational crisis, the population selected for this study is one that is not usually considered to be under threat: high school teachers in Israel.   (489)

The choice of the population of high school teachers in Israel is explained ("Since") in the following way: The researcher's theoretical approach is that job insecurity is not necessarily associated with an organizational crisis; in other words, a feeling of job insecurity can also be found among a group that is not normally threatened by an organizational crisis. That is to say: The choice of a particular research population (a non-threatened population) is presented by the researchers as justified by means of the theoretical approach that they believe in. From this, it may be implied that their study naturally emerges from a particular theoretical approach. This logical relationship between the theoretical approach and the planning of the study conforms to the structure of the scientific paper, which tends to describe the theory as the point of departure from which the research question and methods are derived. But, it is entirely possible that in the current case, the reasons for the choice of this population were incidental or practical, for example, because the researchers had easy access to this particular population. As Swales explains: "This

---

3.   All the examples in this chapter and in Chapter 4 are taken from the Hebrew corpus (see Section 1.1 above) and translated into English for the purpose of this study. The numbers in parenthesis represent the page numbers in the original corpus.

reversal of the research dynamic is *in its context* neither deceitful nor misrepresentative" (1990:118). On the contrary, it is merely an illustration of a process that is typical of scientific writing: The authors use the logical structure of the paper to reformulate their arguments in a convincing manner, and consequently, this structure can be thought of as a method of persuasion on its own.

Bazerman (1988), in analyzing Compton's paper (see above), also addresses the persuasive power of its structure:

> The reasons why the audience might want to believe the article are imbedded in the article's structure. A representation of the literature establishing and positioning a problem, an accurate understanding of existing knowledge, the drawing of a question sharply, the appropriateness of the research design, the fit of the results – these are what convince […].   (Bazerman 1988:219)

Hunston (1994) suggests a textual analysis of a research paper that demonstrates how each part of it may include a specific persuasive goal. The Introduction persuades the reader that "the research undertaken is necessary and worthwhile, on the grounds that there exist some gaps in knowledge on a topic which is important"; the Method section persuades the reader that "the research was well done, specifically that the subjects represented the groups they were intended to represent and the experimental method avoided distortion"; the Results section persuades the reader that "the statistical packages used were useful and informative"; and the Discussion persuades the reader that "the results make sense and fit with other examples of research, leading to a consistent body of knowledge" (p. 193).

An especially detailed analysis from this perspective has been done by Swales (1990) who examined the introductory chapters of research articles. According to Swales, the purpose of the author in this chapter is "to create a research space." The analysis describes three "moves" that together lead to the creation of a research space, while each one of them may employ several options:

1. Establishing a territory – by claiming centrality, interest or importance, making a topic generalization, reference to previous literature etc.
2. Establishing a niche – by indicating a gap, raising a question or a problem, declaring that the research continues a tradition etc.
3. Occupying the niche – by indicating the main purpose of the research, providing the main findings, indicating the structure of the article.

Swales' CARS (= Create A Research Space) model reflects "the need to re-establish in the eyes of the discourse community the significance of the research field itself; the need to 'situate' the actual research in terms of that significance; and the need to show how this niche in the wider ecosystem will be occupied and defended" (Swales 1990:142).

The structure of the academic paper thus seems to be a promising point of departure for a rhetorical discussion. Linguists and discourse analysts have used the IMRD scheme since the 1970s to examine the linguistic character of each part of the paper and compare them (see an overview in Swales 1990: 130–137, an overview in Lewin et al. 2001: 19–20; see also Dudley-Evans 1989; Hunston 1994; Brett 1994; Hyland 1998; Bondi 2004; Bondi et al. 2004; Lindeberg 2004). Many recent studies have analyzed the various parts of the paper in different disciplines and languages in term of argumentative "moves."

However, not all the papers in all disciplines follow this schematic structure. In my corpus of Social Sciences papers in Hebrew, four of the 30 papers do not contain the structural category 'Discussion', eight papers do not contain a 'Methodology' section and eight papers do not contain a 'Results' section. In the humanities, the structure is even more open to variations. Consequently, in the next section I will propose an analysis of the persuasive goals of the academic paper unrelated to its structure, assuming that indications of these goals may be spread throughout various parts of it.

3.2 The persuasive goals of research articles

Following the ideas presented in this chapter, several specific argumentative aims that the author must attain through language may be identified and noted:

a.   To convince the reader that the subject of the research is important and of major significance;
b.   To prove that although the subject has been researched, there are still gaps in knowledge regarding it, i.e. that there remains a research niche, one that is neither too minor nor marginal;
c.   To justify the choice of the theoretical or conceptual framework and of the research methodology;
d.   To convince the reader that the conclusions that the researcher infers from the findings are valid;
e.   To prove that the results make sense and fit in with previous research, leading to a consistent body of knowledge;
f.   To convince the reader that the results represent an innovation.

In two places, it is necessary to balance two of the aims. First, there is an inherent tension between the first two goals. In order to convince the reader of the centrality of the subject, a review of what has already been written about it is usually required. But this review may create the impression that the subject has already been exhausted and that there is no need for any further research or discussion about it. On the other hand, focusing attention on what has not yet been studied may give the impression that the reason these subjects have not been studied is that they are not important. The importance of this point will be demonstrated in Section 4.3.2 by means of the concession structure.

Secondly, there is a certain implied tension between the last two goals as well. Every scientific article should contain an innovation, but in order to gain acceptance, the innovation should not be too sizeable. It should fit in somehow with what is already known. Thomas Kuhn, in his book *The Structure of Scientific Revolutions* (1962) already extensively discussed the limitations of the "innovation" that might easily be accepted by the members of the discipline. The nature of what he called "normal science" is that the bulk of the scientific work is not directed at revolutionary innovations, but rather at the refinement of the formulation of phenomena and theories that are already part of the accepted paradigm of the research community. From this, it follows that the "innovation" that we are discussing here is relatively small, and in many senses is a continuation of accepted traditions of research and thinking. Nevertheless, the term "innovation" is a prevalent one in the context of judging a paper when it is being reviewed for publication in a scientific journal, and the authors of scientific articles put considerable effort into underscoring their papers' "news value" (Berkenkotter and Huckin 1995). Hyland (1999) articulated this tension very aptly: The author of the scientific paper has to convince the reader that his or her argument "is both novel and sound" (p. 359).

Here are a number of examples from my corpus that explicitly relate to each of the aims enumerated above.

(a) *Importance of the subject.* As noted, the first stage on the way to establishing the territory is to present the subject as important and interesting. Utterances that state this explicitly may appear in the paper's opening sentence, as in Example (2).

> (2)   Adjustment following an intercultural transition is *one of the main subjects* studied in the research of the psychology of migration [...], and represents *a central area of interest* in the study of the immigration from the Soviet Union from the former Soviet Union [...].   (199)

However, claims regarding the importance of the subject or the degree of interest in it may be distributed throughout the paper, as in Examples (3)–(4).

> (3)   The potential of the ethnic divide flaring up in Israel is *certainly worthy of further debate.*   (8)

(4)   These organizations present researchers and organizational consultants with *riveting* intellectual challenges, both theoretical and practical.   (530)

Towards the end of the paper, we may find statements explaining that the subject is important because it has practical applications and because the new knowledge on the subject can serve other scholars in the future. The three following examples are the final sentences of papers:

(5)   There can be no doubt that familiarization with myths from the past *can serve* as a warning to professionals in the present too (see Hare 1962).   (546)

(6)   Identifying the various mechanisms that affect the scope of the job [...] *may contribute* to the success of these or other changes.   (655)

(7)   Finally, we would like to reiterate that the socioeconomic status index of occupations *provides important information* on the stratified structure of Israeli society. *It is our hope that it will serve* the community of researchers in the social sciences.   (716)

(b) *Establishing the niche.* Another stage on the way to establishing the research expanse is to describe the niche that the specific study fills. Two matters require justification here: One is the very existence of the niche and the other is its importance. These statements describe the existence of the niche on the background of the gaps in current knowledge. The gaps are often described by means of sentences that contain an explicit negative.

(8)   The current study [...] deals with a subject *that has not been studied at all* in educational research.   (655)

(9)   To the best of my knowledge, *no research has been carried out* to explore the effect of metacognitive guidance [...].   (664)

(10) The studies described above *did not examine* the connection between the cultural identity of the participants in the Naaleh 16 program and their level of adjustment. *This level is examined in the current study.*   (24)

Other striking statements argue the importance of the study by presenting it as the first of its kind:

(11) The survey presented here is *the first attempt* to explore the direct attitude of the Israeli public to grief [...].   (281)

(12) The current study represents *the first link* in the study of the reversal of family roles among adolescents in the normal population.   (436)

(13) In this study, we once again explored the impact of family size on the number of years of schooling, but *for the first time in Israel*, we also examined its effect on IQ, scholastic achievements and stratified ambitions.   (195)

(c) *Justifying the choice of the theoretical framework, the methodology and the research tools.* Every scientific work involves choosing, even if unconsciously, a theoretical or conceptual framework, research method and specific research tools, and sometimes a method of statistical analysis. All these choices can be validated and consequently, the researcher should be able to justify them. In Example (14), the author briefly presents a number of options, and explains his particular choice by stating that the subject of the study was best "suited" to one of them. This claim is explained in the subsequent sentences, which are not cited here:

(14) The potential for the eruption of the ethnic divide can be seen by means of a number of conceptual frameworks, such as that of the "ethnic conflict," the perspective of the "state deterioration," or that of the "collective activity" and "social movements." The current context of ethnic relations in Israel is *suitable, in my opinion, to the choice of the last perspective noted*, that of collective activity and social movements.   (8)

When choosing a research method in the social sciences, the first and most important choice is between qualitative and quantitative research. In the following example, we will see the justification for the choice of a qualitative research method:

(15) The essential nature of the qualitative research approach is to try to uncover internal processes – beliefs, thoughts, emotions and intentions. This type of study makes it possible to explain a phenomenon and experience from the personal and varied perspectives of the participants themselves […]. *This approach best suits our research*: to describe and understand the life experience of teachers when making ethical decisions as well as the behavior they employ to implement them.   (446)

Examples (16)–(17) contain justifications of the choice of specific research tools:

(16) *The test was found to be suitable* for the research for the following reasons: […].   (43)

(17) The following assessment indices were chosen *because they were more successful* than previous indices of the assessment by teachers and friends […].   (83)

(d) *Justifying the results' interpretation and the conclusions.* In my corpus, I found very few utterances that related specifically to these two aims: to convince the reader of the feasibility of interpretation given to the findings and the validity of the conclusions that the researcher seeks to derive from them. These two categories – the interpretation of the findings and the drawing of conclusions from them – are somewhat problematic categories. From a structural standpoint, this is where the conclusion of the entire argument should appear, that conclusion that seeks to become a new scientific fact. What the researcher proposes at these stages is presented for the reader's judgment and consequently, should be able to speak for itself. Because of the unstable status of the utterances in these sections of the paper, it is in them that we find the largest number of expressions of doubt and reservation (Levin et al. 2001:98). In neither of these places do we generally find explicit statements that ask the reader to recognize the correctness of the proposed interpretation or the validity of the conclusions. In fact, the researcher can only hope that the reader will be convinced based on the arguments themselves, the disciplinary knowledge at his disposal and the fact that he is a rational individual.

Moreover, the importance of the subject, the existence of the niche, the coherence between the new and existing knowledge and the innovation that it contains – all these are essentially based on knowledge that goes beyond the area of the particular study. It is impossible to establish the field of research and the niche without first presenting the existing knowledge, and similarly, establishing coherence and innovation must also be done in the context of comparison with the existing knowledge. In order to enable the reader to judge this new knowledge, the researcher must bring to the text the relevant information that is related to it.

On the other hand, providing an interpretation of findings and drawing conclusions from them are actions that are part of the argumentation itself, and should arise "naturally" in the mind of the readers too. While the process of interpretation and drawing of conclusions is also based on disciplinary knowledge and a consensus among the members of the community, it should arise mainly from within the structure of the argument itself. The researcher expresses his interpretation and conclusions, but he is not generally supposed to justify them or argue their correctness. On the contrary, the researcher generally aspires to present his conclusions as arising "on their own" (Livnat 2010a).

Examples (18)–(19) contain utterances that provide support for the correctness of the interpretation or the validity of the conclusion. In both cases, the claim of coherence is used to justify the researcher's interpretation or conclusion. However, it is interesting to note that the justification given in these cases for the choice of a particular interpretation or conclusion is the result of other studies. In other words, we are not talking about bolstering the argument from "the inside," from within the inner process of providing interpretation or drawing conclusions, but from "the outside," in order to justify the internal process.

(18)   According to one interpretation of this finding, during the initial period, the immigrants view the very acquisition of language skills to be a kind of guarantee of their chances [...]. *This interpretation is supported* by the findings of Gittleman's research [...].   (145)

(19)   *Support for this conclusion* can be found in reports from working women who claim that work outside the home is no less important than caring for young children [...]. *A similar tendency was found* in other countries too [...].   (654)

(e) *Coherence and consistency.* Considerable importance is attributed to the new knowledge being perceived as being coherently integrated into the existing knowledge. This integration or compatibility validate the new finding, making it difficult for those that might object to it because the objection could involve coming out against a large body of knowledge that has already been established (Latour and Woolgar 1997: 241). Moreover, presenting new findings as being coherent and consistent with the existing knowledge gives an impression of progress in the entire scientific enterprise in accordance with an orderly and logical program, generally one step at a time. Utterances that underscore coherence could be found in almost every paper in the corpus, for example:

(20)   These findings *coincide* with the conclusions of other studies conducted in the United States.   (276)

(21)   In this sense, the findings are *consistent* with Samoha's assessment (1993) [...]. They *join* a list of research findings showing that the ethnic factor plays a role [...].   (25)

(22)   It may then be noted that the findings of this study *support* the theoretical model and *bolster* empirical findings obtained in other studies.   (505)

(23)   *Other studies also report* on an attachment to the identity of origin and Russian culture.   (584)

(f) *Innovation.* Every academic paper must include an innovation; this need arises from the instructions given to reviewers, in which they are asked to note what innovations the paper has to offer, as well as from the internal discussions held by the editorial board. Towards the end of the paper, in the Discussion, Summary or Conclusion chapters, it is customary to accentuate the innovations in the paper, which are ultimately what the members of the scientific community will be judging. Here are a few examples:

(24) We have further found in this study, *for the first time*, that the children's aspirations regarding social position were negatively correlated with the number of children in the family. (196)

(25) The findings of the study emphasize the importance of the component of fulfillment in the phenomenon of manager burnout, *an element that is rarely seen in the literature*, and may even be at odds with previous findings and conclusions [...]. (238)

(26) *A major contribution* of this study lies in the identification of the differences in responses to job insecurity among different population groups. (505)

(27) Two population groups that *have not received sufficient attention in the research literature*, and that have been identified in this paper are [...]. (508)

In light of this detailed description of the aims that the author seeks to achieve in the academic paper, the hypothesis of my study is that scientific discourse is an argumentative one whose goal is to persuade the scientific community to accept the new knowledge and make it part of the shared, agreed-upon knowledge of that community.

3.3 Degrees of facticity

Viewing the academic paper as a persuasive text directs us to examine the manner in which it handles facts, on the background of a general approach to rhetoric. In order to persuade, the researcher must present his findings as 'facts' through the use of language. The researcher uses language to convince others of the 'facticity' of his findings. In this section, I will provide some insights into the place and status of *facts* in the scientific discourse from the perspectives of argumentation theory, philosophy and sociology of science. Combining these perspectives will present the scientific fact as an entity having both epistemic and social meaning, and the scientific paper as a discourse that on one hand has an epistemic value and role related to knowledge and to the description of the 'world,' while on the other, also has a social value, serving social roles within its relevant discourse community.

Despite its significance, the concept of a *fact* in the "New rhetoric" (Perelman and Olbrechts-Tyteca 1958) is not a developed or elaborated one. P&OT chose to note three basic insights, delivered in two pages of their treatise. In the New Rhetoric, a *fact* is characterized by the agreement of the universal audience.

> In argumentation, the notion of 'fact' is uniquely characterized by the idea that is held of agreements of a certain type relating to certain data [...] The way in

which the universal audience is thought of, and the incarnations of this audience that are recognized, are thus determining factors in deciding what, in a particular case, will be considered to be a fact, characterized by adherence of the universal audience, an adherence such as to require no further strengthening. (P&OT: 67)

While these comments do not specifically refer to scientific facts, in another place in the treatise the authors note that the audience of scientists can be perceived as the universal audience. The scientist addresses himself to certain particularly qualified people, who accept the well-defined system that lies at the basis of their discipline. Yet, this very limited audience is generally considered by the scientist to be the universal audience. He supposes that everyone with the same training, qualifications and information would reach the same conclusions (ibid. 34). Thus, if we take the disciplinary scientific audience to be the universal audience, the ability of a particular piece of knowledge to gain the status of scientific fact is dependent on its acceptance as a fact by the members of the scientific discourse community.

Since the definition of a piece of knowledge as a fact is dependent on the audience's agreement, "factual status" is taken by P&OT to be dynamic, rather than static: "No statement can be assured of definitively enjoying this status, because the agreement can always be called into question later" (ibid.: 67).

According to this approach, the factual status of a particular statement can change, unrelated to the question of whether the reality that it is describing has changed. Thus, it weakens the connection between "facts" and the reflection of reality.

According to P&OT, there are certain conditions favoring the agreement of the universal audience. "A fact serving as premise is an uncontroverted fact. [...] A fact loses its status as soon as it is no longer used as a possible starting point, but as a conclusion of an argumentation" (ibid.: 68). Thus, a change in the discursive status or linguistic form will lead to a change in the "facticity" of that particular piece of knowledge, as it is interpreted by its audience. In other words, language is what creates the facticity, or at least has a profound impact on it.

It is clear, however, and taken as a given in most theories of language, that the connection between language and reality is more complex, and the manner in which the language creates reality should be framed in more subtle terms. A scientific fact is an entity having epistemic value, and there is little point in describing scientific language as creating facts unrelated to physical reality. Language does not only create facticity, but in some respects, it *reflects* facticity, or perhaps reflects the degree of facticity of the various knowledge items in the text. I will return to this point later, in the hope that the discussion will help to produce a more precise understanding.

An interesting perspective that has contributed enormously to the understanding of the nature of "facts" in scientific discourse is provided by Latour and Woolgar (1979) in their book "Laboratory Life," which discusses the social process of creating a scientific fact. Their ideas are based on a close observation of the work of scientists in an endocrinological laboratory, using a methodology adapted from anthropological research. Their description presents the scientific fact not as part of natural reality, but rather as the product of communication, interaction and negotiations, and most especially, as the final result of a disciplinary consensus. In other words, they share the new rhetoric's view that the definition of a piece of knowledge as a fact is based on its acceptance as such by the relevant disciplinary discourse community.

Latour and Woolgar describe how scientists constantly deal with statements: They cite, borrow, enhance, diminish, add modalities, propose new combinations. A scientist can notice how his own assertion is rejected, borrowed, cited, confirmed or dissolved by others. Some assertions change their status as they are proven, disproven, and proven again. In situations where a statement is borrowed, used and reused, there comes a stage when it is no longer contested. At this stage, *a fact has then been constituted*. This is a comparatively rare event, but when it occurs, a statement becomes incorporated into the stock of taken-for-granted features which silently disappear from the conscious concerns of daily scientific activity (Latour and Woolgar 1979: 86–87). If we follow the history of an isolated "fact," as did Latour and Woolgar, we can see how propositions that at a specific time, have not yet earned the status of facts can attain that status at a later time. They can also move in the opposite direction: What once, in the past, attained the status of being considered a fact, can later lose that status.

A similar attitude, one that ascribes a major role to communication and agreement, is expressed by Pera (1994) when discussing the connection between objects, facts and reality from the perspective of the philosophy of science:

> Let "the sun has spots" be a private observational report of an individual *I*. Once he has made it, *I* brings this report to the linguistic market of experts in physics and astronomy where he encounters both buyers and critics. A debate ensues. If the debate favors him – for example, if, after an exchange of arguments and counter-arguments, everybody answers yes or reacts in the same way to the question, "Are there spots on the sun?" or to the command, "Look at the spots on the sun!" – then one can say that sunspots are an object and that the sun has spots is a fact. (p. 159)

Latour and Woolgar (1979) consequently identified five different "statement types" in the scientific discourse (ibid.: 75–79), creating a sort of a scale of "degrees of facticity," from the most fact-like entities to the least fact-like ones. Although this scale is not free of shortcomings – a state that the authors themselves were aware of – the very recognition of the existence of such a continuum has great importance, since it sidesteps the problematic division of statements into only two categories: "fact" or "non-fact."

Following Latour and Woolgar, in this section, I will propose a categorization of the types of knowledge present in scientific papers from the point of view of their facticity, and discuss some of their discursive and linguistic properties. The classification is not exhaustive and the list may be incomplete, but it will neverthe-less help us to clarify the concept of "degree of facticity."

(a) *Implicit accepted, shared knowledge.* In this category, we find unspoken items of knowledge representing taken-for-granted facts, accepted knowledge shared by the members of the discipline. Let us look, for example, at the opening sentences of a paper entitled, "Gender differences at the start of the careers of business ad-ministration graduates." This example contains three knowledge items presented as basic facts and that serve as the background and point of departure for the argument. (The capital letters in brackets here and in the examples that will follow have been added by me.)

(28) [A] The gender makeup of the labor market in Israel has changed in the past decades. The proportion of working women has risen consistently from 10 percent in 1920 to 45 percent in the mid 1990s (Central Bureau of Statistics, 1995), a proportion similar to that found in other Western coun-tries. [B] A number of factors have influenced this participation of women in the workforce, including the rise in the level of education among women, which has increased their ability to be employed in a variety of professions and roles (Oppenheimer 1970). [C] At present, women represent more than half of the students in Israel's institutions of higher education.   (262)

Let us examine these utterances from the perspective of their facticity. The sen-tences that I have marked as [A] contain knowledge items that are reported based on figures provided by the Central Bureau of Statistics (CBS), a governmental agency. According to the manner in which they are presented in this text, it is clear that their credibility is beyond question. The consensus regarding these data is based on (a) faith in the reliability of the CBS's research methods; (b) faith in the professional integrity and scientific objectivity of the people behind these fig-ures. The fact that the authors did not bother to provide further support for these figures by making these beliefs explicit is indicative of the broad consensus that they enjoy. They are 'black boxes' (Latour 1987), i.e. unquestioned premises. This broad consensus is what makes these data into knowledge items with the highest degree of facticity.

On the other hand, the sentence that I have marked as [B] contains a claim that appears to be based on the finding of another researcher. This researcher, who is mentioned by name, appears to be the source upon which the facticity of this knowledge item is based. However, it should be noted that the specific study that is named conducted its research on this subject in the United States, rather than in Israel. Consequently, the claim that appears in the paper regarding the factors that have affected the situation in Israel is a claim made only by the authors themselves. This claim is based on the implicit assumption that there is no reason to believe that the factors that influenced the increase in Israel are different from those in the United States. This assumption is based on yet another implicit assumption, which is embodied in the phrase "other Western countries" at the end of statement [A]. The adjective 'other', which relates anaphorically to Israel, is indicative of the assumption that Israel is a "Western" country. Based on another implicit assumption, according to which all "similar" countries (the "Western" ones, in this case) can be expected to undergo similar social processes, another assumption is derived: These social processes in Israel are likely to resemble those in the United States. The fact that these assumptions are implicit presents them as being consensual, because otherwise they would have to be presented as explicit claims supported by some form of data.

In this sense, rather than being a syllogism, the scientific argument presented here is an enthymeme, some of whose elements have not been revealed. Unlike a syllogism, the enthymeme does not have to explicitly mention arguments the truth of which is accepted by both the speaker and his audience, or present all the steps that led to the conclusion. This kind of implicit knowledge can also be discussed in terms of *doxa* (Amossy 2002) or *topos* (Anscombre 1995), which are unspoken elements, unexpressed shared premises or "beliefs presented as common to a given collectivity" (Anscombre 1995: 35, translated in Amossy 2002: 386). For Anscombre, topoï are:

> general principles on which reasoning relies but which are themselves not reasoning. They are never asserted in the sense that the speaker does not present himself as the author of these principles (even if he actually is), but they are made use of. They are almost always presented as being an object of consensus in a more or less large community.   (Anscombre 1995: 35)

These elements enable the linkage of arguments and allow for effective interpretation and interaction, because they have the "capacity to unveil the hidden principles of argumentative orientation, without which neither meaning nor utterance linkage could be accounted for" (Amossy 2002: 387).

(b) *Explicit accepted knowledge.* In this category, we find utterances that represent facts over which there is apparently no disagreement. Included here, for example, are the findings of others that appear in the text without mention of their authority, i.e. without attributing them to someone else. Their appearance without any specific attribution gives them a high facticity status.

According to Latour and Woolgar, one of the textual indications that signal that a fact has been created is the disappearance of traces of the circumstances surrounding its inception, for example that it stops being attributed to a specific author. "By noting that human agency was involved in its production, the inclusion of a reference diminishes the likelihood that the statement will be accepted as an 'objective fact of nature'" (Latour and Woolgar 1979: 80).

Latour and Woolgar have found that although this type of statement is often taken as the prototype of a scientific assertion, it is actually quite rare in research papers (ibid.: 77). The reason for its infrequency is the convention of scientific writing that requires the mention of an authority for every knowledge item whenever possible. Nevertheless, an example of this type could be the sentence that I have marked as [C] in Example (28) above. Here, there is no addition of the mention of the authority because the item is easily and simply available, one for which it is difficult to imagine that there could be any disagreement. The question of why items such as this should at all appear in a scientific paper can be answered in terms of *presence.*

> By the very fact of selecting certain elements and presenting them to the audience, their importance and pertinence to the discussion is implied. […]. *Presence* […] is an essential factor in argumentation and one that is far too much neglected in rationalistic conceptions of reasoning.   (P&OT: 116)

In the case of Example (28), this information is closely connected to the structure of the argument that will be presented later in the paper, and consequently must be made present in the text in an earlier stage.

(c) *Other's findings or results.* This category refers to findings made by others who are mentioned by name, that is, utterances that attribute the claim to someone other than the author. What is the degree of facticity of items that are explicitly attributed to someone other than the author? From the previous section, it would appear that utterances that are not explicitly attributed to anyone have a higher degree of facticity, since they are perceived as common and consensual knowledge. From this, it follows that findings that are attributed to authors could have a lower degree of facticity. Myers (1989) stresses that "any attribution of a statement to a person weakens it" (p. 6). On the other hand, it could also be argued that "the inclusion of a reference lends weight to a statement which otherwise appears to be an unsupported assertion. Thus, it is only by virtue of the reference that the statement achieves any degree of facticity" (Latour and Woolgar 1979:80). In other words, attributing the claim to someone other than the author, noting the source of the claim, or even noting the very fact that it has merited scientific publication, grants the statement the status of being accepted as disciplinary knowledge. It is thus difficult to say whether the noting of an authority, namely, attributing the statement to another author, strengthens or weakens the factual status of the utterance.

Another question is the manner in which the findings of previous scientific studies are integrated into the paper's argument. Other's findings are occasionally presented in order to support the conclusion that the writer seeks to draw, or the interpretation that he is offering for his data, as in Example (29).

> (29)   [A] We can say that at the early stage, the language fills only a symbolic role […]. As time passes, the symbolic role diminishes. [B] *According to one interpretation of this finding*, in the initial period, the immigrants view the acquisition of language skills as a kind of promise. […] [C] *This interpretation is supported* by the findings of the study by Gitelman (1995), who found that […].   (145)

What I have marked as ⌊A⌋ represents the findings of the researcher himself; ⌊B⌋ represents the researcher's own interpretation of these findings; and [C] represents the findings of a previous study (= Gitelman 1995), which are presented here in order to support the interpretation that the researcher is proposing.

Gitelman's findings appear to be given a high level of facticity. Not only have they already merited scientific publication, but they are presented as information from which one can derive a particular interpretation of the researcher's findings. As mentioned above, following P&OT, one of the ways to strengthen the factual status of an utterance is to use it as the premise of an argument. It can be argued that the very consideration of this information when drawing the new conclusions of the current study, that is, *positioning it as a premise*, helps to raise its factual value. It signals the reader that for the writer, it is a fact.

(d) *Author's findings or results.* The study's findings generally merit a separate section. In this section, the impersonal, "reporting" nature of the language is conspicuous, and its purpose is to present the findings as having been attained by means of "objective," cautious and precise scientific work. In Hebrew, this language is typically characterized by extensive use of passive forms and "grammatical metaphors" (Halliday 2004; Livnat 2010a), as in Examples (30)–(32).

> (30)   […] *this paper proposes* a different perspective […].   (587)

(31) *The discussion focuses* mainly on the behaviors presented [...].   (531)

(32) *The current analysis* [...] *explores* this question [...].   (58)

The impersonal nature of the language increases the chances that the findings will be perceived as facts and their factual value raised. In Example (33) below, the findings themselves, or their interpretation, are presented as a fact simply by use of the word "fact" (Hebrew: *uvda*).

(33) The findings of the study show that [...] groups have demonstrated a higher level of inter-group aggression than the individuals in all situations and according to all indices. This *fact* is further bolstered by the historic impressions of Le Bon (Le Bon 1879) regarding the crowd being dangerous and violent [...].   (410)

What the word 'fact' actually refers to is that in the eyes of the author, it is a fact. However, it is important to stress that at this stage, the findings are a fact only in the eyes of the researcher that found them; they are not yet a fact in the eyes of the scientific community. It is a piece of knowledge that has not yet earned the acceptance of the relevant disciplinary community. Nevertheless, the use of the word "fact" may serve to strengthen, at least on the surface, the factual status of the findings.

(e) *Interpretation.* Data never appear in a scientific paper without an explanation provided by the author. From the perspective of argumentation, even when the data are numerical, the text that accompanies them should not be viewed as a mere verbal repetition of the data, but rather as a form of interpretation. The utilization of data for argumentative purposes is impossible without a conceptual development, which gives them meaning and makes them relevant to the progression of the discourse (P&OT: 120).

  While the findings of the study do not yet fall into the category of a scientific fact, the factual status of the interpretation or the conclusion derived thereof is even weaker. Interpretation and conclusion are indirect, deduced evidence, and consequently, their degree of facticity is relatively low. The following example shows the extent to which the interpretation of findings is a matter of free choice:

(34) The connection [...] is indicative of the positive contribution of the preservation of cultural identities [...]. *Nevertheless*, we are aware of the *possibility that there may be additional or alternative causal interpretations* of the findings. *It is possible that* a high level of adaptation is what enables the immigrant [...]. Similarly, *it is possible that* both the adaptation and the multiplicity of identities are affected by another factor. This factor *could be* the duration of time [...].   (213)

This example evinces the freedom that the researchers take in choosing the interpretation that suits their findings. This freedom naturally invites the expression of an epistemic stance and the appearance of modal expressions (Nuyts 2001) and hedges (Hyland 1998), such as "it is possible that," and "could be," as well as concession structures ("nevertheless").

(f) *Conclusions.* Utterances in this category also tend to include what Latour and Woolgar have called "modalities" – a reservation that is expressed by the speaker about his own claims. Lewin et al. (2001) found in Social Sciences papers in English that 50% of the claims that were marked by the authors as 'conclusions' were hedged. Many of the other 50%, which were found to be unmitigated, were previous author's conclusions or claims that represented no challenge to the scientific community (Lewin et al. 2001:97–98). In Example (35), which is taken from a Conclusions section, the modal expression *can* weakens the factual status of the conclusion.

(35)   We have found that [A] the proportion of men filling senior positions is higher than the proportion of women already at the beginning of their careers. Thus, *the differences* that were found *can* be indicative of [B] discriminatory practices in the labor market.   (277)

On the other hand, the facticity of the conclusion is strengthened by virtue of its compatibility with the findings of others. [B] is knowledge that has already been put forth in the past, but it is concluded once again from the findings of the current study. This kind of coherence between findings serves to support the conclusion.

The noun 'differences' is also of interest. It is an example of nominalization – the use of abstract nouns to refer to "virtual entities" (Halliday 2004) – since "a difference" is not really an entity, but a relationship. Nominalization was identified by Halliday (2004:Ch. 4) as a clear linguistic feature of scientific discourse. The first sentence of Example (35) describes these differences, and in the second sentence, these differences are given a definite noun form. Using the definite noun and placing it in a thematic role bolsters the factual status of those differences. In contrast, the use of the noun phrase "discriminatory practices in the labor market" without a definite article weakens the factual status of the conclusion.

(g) *Conjectures and speculations.* These optional elements appear most commonly at the end of papers. Let us look at an example from a paper that discusses virtual communities. The heading of the final section in the paper is: "Future influence of virtual communities on Israeli society." This section contains certain predictions for the future:

(36) *If* the virtual communities become [= Hebrew: *will become*] a significant
      factor in the life of the individual in Israel, they *could* slow down the rate of
      settlement in the center, and *perhaps* even cause people that live in the center
      to move to the periphery [...].   (313–314)

Linguistically, this sentence contains many expressions and structures of the category usually called "irrealis": condition (*if*), future tense (*will become*), modals
(*could*) and hedges (*perhaps*). All these linguistic elements join together to form
utterances that have a very low degree of facticity. In fact, in this case, no commitment to the truth is required. What is presented needs only to be an intelligent
and rationally established guess.

The above discussion provides some insight into the connection between
scientific language and facts: Does scientific language *reflect* facticity or does it
create it?

In what sense can it be said that the language *reflects* facticity? Certainly not
in the simplest sense, according to which the fact is somewhere "out there" in
the world of reality, and the researcher that discovered it needs only to put it in
writing. Indeed, language in essence has this nature, but in the scientific context, its chances of doing so successfully are slim. It would be more correct to say
that the scientific language reflects the *degree* of facticity of the utterance at every
stage: the manner in which each piece of knowledge is formulated, the addition
or omission of modalities and hedges, applying authority, namely the addition of
reference, the positioning of the utterance as a premise or conclusion of an argument – all these and other linguistic and discursive elements, reflect the current
status of the utterance, the degree of acceptance that already exists in relation to
it, the extent to which it is already perceived by the scientific community as a *fact*,
the distance that it still has to cover until it becomes a taken-for-granted fact. This
status is dynamic and open to discussion, and it is presented from the author's
point of view. In the future texts, both written by the author and by others, a different status can be reflected.

In what sense can it be said that the language *creates* facticity? In the scientific context, language creates facts in the sense of social acceptance. By using
language, the researcher *reports on his belief* that what he is describing is indeed
a fact. He hopes to transmit that same belief to his audience, thereby gaining its
acceptance. His expectation to gain acceptance is based on the nature of the universal audience and its rationality. Perelman (1982) describes the universal audience as consisting of "all of humanity, or at least all those who are competent
and reasonable" (Perelman 1982: 14). In the context of an audience of scientists,
rationality as a general quality can be replaced with what we may call *disciplinary rationality*. In choosing the term "disciplinary rationality" rather than a more

general one "scientific rationality" I address the fact that questions such as how a proper scientific study should be conducted, what method is considered reliable, what kind of evidence is proper evidence and what should be considered proof vary from one scientific discipline to another. In this sense, the disciplinary community may be considered as a particular, rather than a universal, audience. Disciplinary rationality includes shared knowledge and acceptance of disciplinary norms. Thanks to these accepted norms, which are shared by both the researcher and audience, the researcher hopes that the audience will be convinced that his study was conducted properly, and that knowledge worthy of being considered 'a fact' was attained. In this way, he hopes to pass on to that audience the same belief in the facticity of the knowledge, a belief that he himself holds.

In fact, this is the essence of the scientific paper as an argumentative text. Its point of departure is *the faith of the speaker* in the facticity of the new information, and its rhetorical aim is to convince the audience, based on its disciplinary rationality, to accept that same faith and thereby usher the new claim into the shared disciplinary body of knowledge. Beller (1999) describes the rhetoric used by the physicist Niels Bohr in some of his papers on the quantum theory, as a "strategy of persuasion, which infects the reader with the sheer intensity of Bohr's conviction" (Beller 1999: 192–193).

3.4 Argumentation, facticity, time: Three parallel lines

Following the discussion in the previous section, it is now possible to propose that generally speaking, the scientific paper moves simultaneously along three parallel lines: The line of argument, the line of facticity and the line of time. This parallelism is illustrated in the following sketch:

Premises	$\longrightarrow$	Conclusions
Facts	$\longrightarrow$	Non-facts
Past	$\longrightarrow$	Future

As with any argumentation, the logical and argumentative structure of the scientific paper moves from the premises to the conclusions (Van Eemeren et al. 2002) or from the Data to the Claim (Toulmin 1958). The combination of the existing, accepted knowledge together with the findings of the current study creates the premises, or the available data, from which the conclusions are derived, which design a new knowledge claim. According to the writing conventions practiced in modern journals, the place of the conclusions or the new claims is towards the end of the paper, after the premises or the relevant data that support them have been properly presented.

As we have seen, the structure of the scientific paper moves also from the most fact-like entities to the least fact-like ones. The paper usually begins with a survey of the shared disciplinary knowledge, which has the highest level of facticity. It continues with the findings of the study, the degree of whose facticity remains open: Although they are presented by the researcher as facts, they have still not earned the recognition of the disciplinary community as facts. From here, it moves on to interpretation and conclusions, whose degree of facticity is even lower, and the paper sometimes concludes with statements aimed at the future, which do not have even a semblance of facticity.

Interestingly, these two lines converge with another line that the paper moves along: the line of time. The structure of the scientific paper typically moves from the past to the future. The shared disciplinary knowledge is oriented to the past, since it is based on studies carried out and published before the current study. The current study, its findings and conclusions represent the present, or at least the very recent past. These are frequently joined by utterances aimed at the future: not only predictions and speculations, but also proposals for future research. Dudley-Evans (1989), who described the rhetoric structure of the Discussion/Conclusions chapter, has pointed to the presence of a "move" he called "recommendation for further research." Berkenkotter and Huckin (1995) have suggested that the global structure of the Discussion/Conclusions section is the mirror image of the moves of Introductions. While the Introduction section includes a move that Swales (1990) called "establishing a territory" (see Section 3.1 above), the Discussion/Conclusions section may include a move of "establishing additional territory" that the authors can perform by showing the implications of their contribution (see also Lindeberg 2004: 43–44). This move by nature points forward, to the future.

While not all papers answer the description that I propose here, it nevertheless appears to be crucial for our understanding of the unique nature of the discourse of scientific papers.

The discussion in the current chapter clearly points to the argumentative nature of scientific papers. Assuming that argumentation is necessarily dialogic, we may argue that dialogicity, as presented in Chapter 2 above, has specific features that pertain to its persuasive function. I call these features the "dialogic dimension" of academic discourse, and will address this dimension in the next chapter by discussing four linguistic structures: citations, concession, *we* forms and questions.

CHAPTER 4

The dialogic dimension of academic discourse

4.1 Towards a new model of scientific dialogicity

When discussing speech genres, Bakhtin distinguishes between primary, simple genres, such as everyday conversation, and secondary genres, such as the novel or scientific discourse. Secondary genres have a unique dialogic quality since they are more developed, complex and pre-organized. Secondary speech genres by their very nature introduce primary genres and relations among them into the construction of the utterance (Bakhtin 1986: 72–73). A unique kind of responsive understanding is characteristic of secondary genres – a silent responsive understanding with a delayed action (ibid.: 68). The response of the reader may arrive after a long time, in a different time and place.

The typical dialogicity of the academic discourse can be examined along the two dimensions of intertextuality: the horizontal and the vertical (see Section 2.1 above). As for the horizontal dimension that connects the text to other past and future texts, the significant fact is that the scientific paper is part of a larger "scientific project," one link in a chain of texts, each of which has its own unique role in this project. Here the *manifest intertextuality* is evident.

However, in academic discourse, manifest intertextuality is expressed not only by citing others, but also via many different textual elements. According to the approach that I propose here, and following the theories that have been presented in the previous chapter, dialogism means not only an appeal to other texts but also to the audience, and in the case of written texts – to the reader too.

The question of the "audience" of the academic paper is a complex one. To whom is the text directed, and how can an investigation of its textual design help us answer this question?

From the perspective of politeness theory, Myers (1989: 3) suggests that we should distinguish between two different audiences: (1) the wider scientific community and (2) an immediate audience of individual researchers and particular groups of researchers doing similar work. Myers mentions that there are many other possible audiences: science journalists, administrators, biographers, sociologists of science etc. However,

> [...] the only audiences that matter here are those needed to account for features of the text. So while it is true that some sentences from these articles [= molecular genetics articles] are about to be read by linguists, it is not necessary to include these linguists as one of the potential audiences, because I cannot find any features that can be explained by taking this audience into account.
>
> (Myers 1989: 3–4)

The relevant audience is thus the one that the writer directs his writing at, and this directionality is generally expressed through specific components of the texts.

During the last decade, two quite detailed descriptions of the polyphony of academic papers and the interaction between the various voices in it have been proposed, both corpus-based studies.

Ken Hyland (2001a, 2005) based his research on a corpus comprised of 240 research articles in English, in eight disciplines: mechanical engineering, electric engineering, marketing, philosophy, sociology, applied linguistics, physics and microbiology.

A similar project was taken up by three other researchers, Kjersti Fløttum, Trine Dahl and Torodd Kinn (2006) based on the KIAP corpus, which consisted of 450 research articles from three disciplines: economics, linguistics and medicine, and in three languages: English, French and Norwegian.

These two models include explicit features of the writer's stance and positioning, on the one hand, and explicit features of reader orientation, providing evidence of reader engagement in the text, on the other. Table 1 presents the categories of linguistic components that were examined in these two projects.

Let me briefly clarify and exemplify some of these categories:

Hyland's "self-mention" refers to the use of first-person pronouns and possessive adjectives. The category "attitude markers" includes verbs such as *agree, prefer*, sentence adverbs such as *unfortunately, hopefully*, and adjectives such as *appropriate, logical, remarkable*. "Hedges" are devices such as *possible, might, perhaps*; "Boosters" are words like *clearly, obviously, demonstrate.*

By "personal asides," Hyland is referring to interruptions in the ongoing discourse, briefly breaking off the argument to offer a meta-comment on an aspect of what has been said (Hyland 2001a: 561).

The category "indefinite pronouns" in Fløttum et al.'s project includes the English *one*, the French *on* and the Norwegian *man, en/ein*. In the category of "meta-text" they include nouns such as *article, paper, (sub)section*, and deictic markers such as *above, below* and *now*.

All these elements are expressions of *manifest intertextuality* that is related to the presence of other texts and of the reader in the written text. However, scientific dialogicity is also of the kind that Fairclough (1992) called *constitutive intertextuality*, one that is related to the adoption and assimilation of genre conventions.

Table 1. Elements of scientific dialogicity

	Hyland (2001a, 2005)	Fløttum et al. (2006)
Writer stance and positioning	Self-mention Attitude markers Hedges Boosters	Pronouns (+ certain verbs) Evaluative elements ('comprehensive', 'fruitful', 'important')
Reader engagement, Reader/writer interaction	Inclusive first-person pronouns ('one', 'the reader') Imperatives ('note', 'consider') Personal asides Questions (real and rhetorical) Obligation modals ('must') 'It is (adjective) to do' Appeals to shared knowledge	Inclusive first-person pronouns Indefinite pronouns 'Let-us' imperatives Metatext
Others		Indefinite pronouns Negation Adversatives Concession Bibliographical references

The accepted norms of writing within the discourse community, such as those that are embedded in the academic discourse, express the voice of the community. The fact that the text is a link in a chain of texts and that its wording is the result of exposure to previous texts is "manifested in the overtones of the style, in the finest nuances of the composition" (Bakhtin 1986:92). Only the adoption and use of the disciplinary genre conventions enable the writer to create a dialogue with the disciplinary community, to get the floor and the permission to voice his claims. Controlling the shared lexicon is also a major means of drawing the boundaries between those who are included in the discourse community and those who are not, along with the use of acronyms, preferred metaphors, familiar argument structures, citing practices, all of which "foreground a common frame for seeing the world, identifying problems, and solving issues" (Hyland 2001a:566). In these cases, the dialogic dimension is implicit and is embedded in the style of writing.

This aspect of dialogicity is intimately connected to the author's effort to position himself as a member in good standing of the disciplinary community. As Thompson (1996) noted, "academic communities recognize certain wordings as belonging to a particular theoretical orientation. This kind of voicing can clearly serve a solidarity function, being intended to be recognized only by those within the community" (Thompson 1996:510).

Furthermore, not only are linguistic and textual norms embedded in the academic writing, but also shared accepted knowledge, beliefs and attitudes, ideological and epistemological stances (Silver 2004), agreements in regard to methodology, objectivity, ethics and more. Hyland (2001a, 2005) refers to a dialogic component that he calls "appeal to shared knowledge," defined as "the presence of explicit markers where readers are asked to recognize something as familiar or accepted" (2005: 184). This component is more frequent in the "soft" sciences, and in English is often realized through the adverbial phrase 'of course' (Bondi 2002).

These elements of constitutive intertextuality are thus related to the vertical dimension of dialogicity, since they connect the text with other texts that are more or less contemporary with it. They are involved in creating solidarity with the community through the appeal to its shared topoï (see Section 3.3 above).

Among the large variety of textual elements, in the present chapter I will discuss only four: citations, concession, the inclusive *we* and questions. Just as none of the other models provides a full picture of the dialogicity of academic papers, the present chapter does not pretend to provide such a picture either, and nor is this its object. These elements were chosen because they are frequent and important components of scientific dialogicity. Their distribution, functions and significance will be demonstrated in the analysis of full papers in Chapter 5 below. I analyze them here in a representative corpus, in order to first create the basis for defining them as dialogic components. My investigations of each of them will also provide some new insights regarding their specific properties and functions in the academic discourse.

The role of concession in academic articles has never received the attention I believe it warrants. It has been mentioned only briefly by Swales (1990: 154), Hyland (2001: 569) and Fløttum et al. (2006: 247–250), in connection with adversatives. Hyland (2009: 39) concedes that it is a highly dialogistic and productive device in persuasive discourse, but gives no detailed analysis of this construction. For that reason I will provide an introduction to explain the potential force of concession as a persuasive device, with many examples of this linguistic and argumentative structure fulfilling various rhetorical roles in research articles (Section 4.3).

Many studies discuss the forms and functions of citations in academic discourse (Becher 1981; Swales 1986, 1990; Jacoby 1987; Bazerman 1984; Hunston 1994; Montgomery 1996; Baynham 1999; Hyland 1999; Lewin et al. 2001; Fløttum 2003; Silver and Bondi 2004; Hunston 2004; Thompson 2005; Fløttum et al. 2006), but the issue of authenticity and responsibility still requires further development (Section 4.2.2). In my discussion of this topic, I will also refer to the problems faced when writing science in a language other than English.

Special attention to the uniqueness of Hebrew is also given in Section 4.4 in the discussion of pronouns, adding some new insights to this extensively discussed topic (see Ivanič 1998; Bazerman 1988; Myers 1989; Bloor 1996; Kuo 1999; Tang and Suganthi 1999; Groom 2000; Hyland 2001a, 2002a, 2005; Samson 2004; Poppi 2004; Fortanet 2004; Harwood 2005; Martín-Martín 2005; Poudat and Loiseau 2005; Fløttum et al. 2006; Čmejrková 2007; Breeze 2010). My discussion in this section suggests a distinction between two types of inclusive *we*, based on their direction of dialogicity. A similar suggestion is provided in Section 4.5 in the discussion of questions, whose functions in academic discourse is also the subject of extensive study (Adams Smith 1987; Swales 1990; Webber 1994; Thompson 1997; Biber et al. 1999; Hyland 2001b, 2002b, 2005).

As a result of the theoretical and conceptual framework and the analysis of examples from the Hebrew corpus, a new model of scientific dialogicity is proposed, one that takes into consideration the multiplex conceptual framework of the present study.

Its schematic form may be drawn as follows:

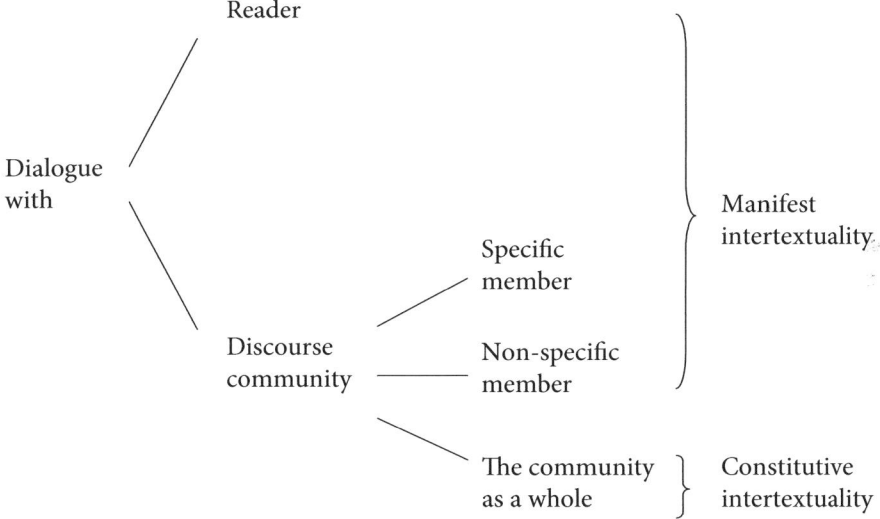

The discussion of the four dialogic elements in this chapter will serve to clarify and justify the distinctions that are proposed in this combined model: dialogue with the reader vs. dialogue with the community, and dialogue with a specific or a non-specific member in the community vs. dialogue with the community as a whole. Towards the end of this chapter, in Section 4.6, I will return to this scheme and rely on the insights that emerge from the discussions to exemplify the different parts of this model.

4.2 Citations

Citations of and references to the research literature are a necessary component and conventional genre requirement in the writing of an academic paper; at the same time they are a major available resource that enables the author to define and establish both the place of the new claim in the pool of disciplinary knowledge and his own place in the disciplinary community. This is the clearest expression of manifest intertextuality, of explicit polyphony (Fløttum et al. 2006: 242). They conduct a dialogue with the discourse community – its members and their shared knowledge.

Tracing the development of the genre of academic papers demonstrates that historically, the number of citations in a paper has grown larger and larger over the years. The reason may be the obvious one: the fact that today we have more items of knowledge and more texts available for citing. But this feature also reflects the intertextual nature of the postmodern age. "The new cultural and creative consciousness lives in an actively polyglot world. The world became polyglot, once and for all and irreversibly," Bakhtin claims (1981: 12). Montgomery (1996) shows how the practice of citing has grown quantitatively over the past century. In 1900, it was common for an article to carry no more than three or four references to other works. By 1950, this had increased substantially; today it is not unusual for reference lists to run two or more pages in length (Montgomery 1996: 39). In my corpus of Social Sciences papers in Hebrew, the average number is 39.3 items per paper.

Generally speaking, the number of citations, the modes of citing and their functions are a matter of the author's choice. Their use, however, is also a function of disciplinary constraints and conventions. Becher (1981: 112) examined how researchers from different disciplines justify the practice of citation. Researchers in biology and physics explained that the cumulative nature of the discipline necessitates a comprehensive initial review of the relevant previous literature in order to ensure that there is no direct replication of previously established results, to establish the niche and to acknowledge academic debts. Historians argued that the literature review may serve mainly to identify the origins of stimulating ideas and issues about which to register dissent. They were conscious of the conventional nature of most annotations and were also willing to admit the role of citations in promoting group solidarity. Sociologists considered citations a useful way of bolstering the argument, but the more cynically minded took the citation of great names as "a vulgar display of knowledge" (p. 112).

In analyzing a paper on the sociology of science, Bazerman (1988) claimed that the reason for the large number of citations at the beginning is that the author needed to establish the ground on which his claim was to rest. The literature of the field does not provide a generally recognized framework in which to place the

Table 2. Number of citations (adopted from Hyland 1999)

Field	Citations per paper
Sociology	104
Philosophy	85.2
Biology	82.7
Applied Linguistics	75.3
Electronic engineering	42.8
Physics	24.8

new claim (p. 34). In the "soft" sciences, he argues, it is sometimes "necessary to confirm to the reader that this topic does exist" (p. 36). Citations may also be considered a politeness device (Myers 1989). Since there are always far more papers that could be cited than the journal would allow, selections have to be made, and thus the decision to cite colleagues might be influenced by the desire to give them the "gift" of credit (Myers 1989: 10–11).

The act of citing is thus a disciplinary act, and can therefore be expected to vary in nature and quantity across disciplines. A comparative study conducted by Hyland (1999) showed that in general, papers in the soft sciences employ more citations than papers in physics or engineering. Table 2 presents some of the findings of Hyland (1999).

These quantitative results can be explained and interpreted using the words of Bazerman (1988) who explains that in social sciences "the literature may be diverse, unsettled and open to interpretations. Therefore, the paper must reconstruct the literature to establish a framework for discussion" (p. 46). Furthermore, "the criteria the audience will apply may not be clear-cut and universal, nor is it certain what intellectual framework they will bring to the reading" (p. 34).

These insights suggest that in the social sciences, and I would add that in the humanities too, the role of citations goes far beyond the simple stating of what is already known about the subject, or what the findings or results of others are. In Hyland's words, we are talking about establishing "a persuasive epistemological and social framework for the acceptance" of the writer's arguments (Hyland 1999: 344).

4.2.1 Patterns of citations

From a structural point of view, Swales (1990) proposed a distinction between integral and non-integral citations. Integral citations are those in which the name of the cited author occurs as a constituent of the citing sentence, as the subject, such as in (1); or in other syntactic positions, as in (2) and (3).

(1)   <u>Brie (1988) showed</u> that the moon is made of cheese.

(2)   The moon's cheesy composition was <u>established by Brie</u> (1988).

(3)   <u>Brie's theory</u> (1988) claims that the moon is made of cheese.

The non-integral citation refers to another author by placing his name in parenthesis or footnoted or endnoted with superscript numbers, as exemplified in Examples (4) and (5).

(4)   It has been shown that the moon is made of cheese (Brie 1988).

(5)   It has been established that the moon is made of cheese.[1]

Fløttum et al. (2006) divided citations into four categories, or four patterns, as shown in (a) to (d):

a.   Non-integral reference: Little Lake is polluted.[1]
b.   Partly integral reference: Little Lake is polluted (Clark 1999).
c.   Semi-integral reference: Clark (1999) has observed that Little Lake is polluted.
d.   Fully-integral reference: (d1) Clark (1999) claims: "Little Lake is polluted."
                                (d2) Clark (1999) claims that "Little Lake is polluted."

In my corpus from the social sciences in Hebrew, I found no examples of pattern (a) at all, apparently due to disciplinary conventions or the requirements of the particular journal. Pattern (d1) was also very rare. Here is one of the few occurrences:

(6)   As Pines (1993) underscored: "Burnout is always the final result of a gradual process of disillusionment, with an eye to eliciting existential meaning from work" (p. 40).   (238)

Example (7) is similar to the previous one, except that the parts of the sentence are reversed, with the subordinate clause following the statement clause:

(7)   However, the problem of representation is deeper and more complex than the superficial argument over the degree of consistency with reality. "What is the 'right' representation of female reality? How can one strive for representation of multiple realities of women's worlds?" asks Lamish (1997, p. 122; see also Ganguly 1992).   (481)

Pattern (d2), in which the words of another researcher are placed in quotation marks within a subordinate clause, appeared only once in the corpus:

(8) Olson has already noted the problematic nature of assuming rationality, and his work, which is largely based on the assumption of rationality, stated as early as 1965 that "The theory developed here [...] is not especially useful for the analysis of groups characterized by a low level of rationality [...]" (Olson 1965, p. 161).   (9)

Example (9) is a combination pattern, since only part of the citation is placed in quotation marks:

(9) Theoretical support for this decision can be found in remarks by Jacobson (1991, pp. 34–35), who maintained that the meaning of the inability to oppose change or threat (helplessness) is identical in the eyes of the worker to the possibility of a change for the worse, "since helplessness in opposing a threat leads to the possibility of a greater loss".   (495)

The most prevalent patterns in the corpus, those that reflect the disciplinary conventions and the requirements of the journal, are patterns (b) and (c). These are also the two principal quotation methods presented by Swales, those that he called the "non-integral citation" and the "integral citation" respectively. Of the 1,200 citations found in my entire corpus, 785 (65.4 percent) are of the pattern (b) type and 384 (32 percent) are of the pattern (c) type. Together they represent 97.4 percent of all the references.

The proportions that were found in the Hebrew corpus are strikingly similar to those that Hyland (1999) found in sociology papers in English (see Table 3). This similarity is not surprising – it merely reflects yet another aspect of the powerful influence that English, as the lingua franca of the academic world, has had on the Hebrew style used in scientific publications.

Table 3. Integral and non-integral citations

	Hebrew – Social Sciences (general)	English – Sociology (Hyland 1999)
Non-integral citations	65.4%	64.6%
Integral citations	32%	35.4%

Because the conventions that determine the choice between an integral and non-integral reference are discipline dependent, as shown in Hyland's study (1999), I have chosen for the purpose of comparison between the languages the disciplines closest to those examined by Hyland. In Hebrew papers from the field of archeology, which I will analyze in Chapter 5, non-integral references take up more than 80 percent. In the case of archeology, this preference would appear to be related to the fact that in these articles, the non-integral reference is frequent because the author is simply noting archeological findings whose interpretation is accepted and consensual. In most cases, the referenced researcher is the scholar who excavated and published his or her finding.

What are the differences between the two main patterns, integral and non-integral citations? One difference is that integral references, especially those in which the cited author's name is in the syntactical position of subject, give more space to the cited *person*, whereas the non-integral references emphasize *the issue* at the expense of the person. A text that includes numerous integral references in the subject position will appear to be organized according to *people* rather than issues. This is illustrated in Example (10):

(10)   There are a number of different views in the research and theoretical literature regarding the areas that should be explored in the study of immigrant adjustment. <u>Taft (1987) maintained that</u> the research of immigrant adjustment should be carried out on two separate dimensions: the internal-personal dimension and the external-social dimension. <u>He suggested that</u> five principal aspects should be investigated on these two dimensions: emotional state, cultural identity, cultural competence, level of social absorption and acquisition of cultural roles. Similarly, <u>Searle and Ward (1990) distinguished between the psychological and socio-cultural levels. On the psychological level, they focused on</u> the individual identity, cultural identity, emotional health and the level of quality of life. On the socio-cultural level, they discussed the aspects of the interaction between the individual and the social institutions and the cultural norms of the target country. On this level, work, family life, education and the acquisition of communication skills are emphasized. <u>Berry (1997) maintained that</u> the distinction between the two levels requires a different treatment for each: On the socio-cultural level, theories that include aspects of social learning should be used, and on the psychological level, theories that include aspects of stress should be used.   (200)

Note that this text is also organized chronologically, and this organization also contributes to the impression of a *narrative* text describing *what people did*. However, this is not the main difference between the patterns, as we shall see below in Section 4.2.2.

In the integral pattern, the name of the cited author appears as a syntactical component of the sentence. According to Fløttum et al. (2006), the most significant and present role another researcher can be given in a text is to be placed in the position of a syntactical subject. Here are a number of such examples from the corpus:

(11)   Oser & Althof (1993) maintained that a debate on the professional dilemma
       [...].   (619)

(12) Kerslake (1981) reached the opposite conclusion [...].   (662)

(13) Alba & Moore (1983) proposed identifying social circles using a two-stage method.   (688)

(14) Duncan [...] sought to assess prestige ratings for occupations by means of two characteristics [...].   (707)

A survey of the verbs that are paired as predicates with this subject will show that not every mention of another author serves to present his arguments. The verb "maintained," as in Example (11), is used explicitly to introduce a claim made by another researcher into the text. The other's argument can also be presented as a conclusion arising from the findings of a study, as in Example (12). The "others" mentioned in this syntactical structure do not only "maintain," but also take other scientific action, for example, they offer methods and models (Examples (13) and (14)). In each of these cases, the responsibility for the claims, conclusions and suggestions is attributed to the other cited, rather than to the author of the paper. This is the typical situation for the integral pattern. The responsibility of the other for the claim can also be presented from a syntactical position that is not the syntactical subject, for example:

(15) According to Gravemeijer (1997), situational knowledge causes barriers that prevent transfer.   (677)

(16) Similarly, if we accept the claim made by Guilford (1950) regarding the continuity of the capabilities of intelligence and creativity [...].   (50)

Sometimes another author is mentioned without any specific claim made by him being clearly presented in the text:

(17) These findings are consistent with the findings of Liran-Alpher (1994), who analyzed the media coverage of women Knesset members.   (469)

What from the cited text is brought into the current text? The author may introduce explicit statements made by another into the text, as in the following example:

(18) The cognitive development is not development that the student arrives at on his own, but rather something he achieves by means of interaction with others: "What a child can do today in cooperation, tomorrow he will be able to do on his own" (Vygotsky 1962, p. 87).   (676)

The quote is a kind of illustration for a claim that has already been made by the speaker, and bringing it into the text is apparently motivated by the fact that the other author articulated it especially convincingly. Consequently, the colon can be understood as a marking of an appositional relationship between the clauses. In addition, the reason for citing Vygotsky may be twofold, both because this is an eloquent articulation of the concept as well as because he is familiar or is likely to be familiar to the readers. The same is true for the citation from Geertz in Example (19):

(19)   Through the ceremony, the individuals explain the essence of the social order in which they find themselves to themselves, as Geertz famously put it – in his commentary on the cockfights in Bali – that the event or ceremony is "the story they tell themselves about themselves" (Geertz 1990, p. 301). The participants in the social event are not always aware of the anthropological interpretation, but the researcher saw in the event a text that he was "strain[ing] to read over the shoulder of those to whom they properly belong" (Geertz 1990, p. 304).   (108)

Further content often transferred verbatim to the text includes definitions that have their source in another, sometimes placed between quotation marks with the specific source of the citation noted, as in Example (20), and sometimes without quotation marks, as in Example (21). In these, there is a possibility that the definition as it appears in the cited source is not identical to the definition in the text.

(20)   The following definition is acceptable: "A ceremony is an act of an individual or individuals, which is essentially symbolic content embodied in action (Wallis 1983, pp. 334–335).   (104)

(21)   A network is a group of individuals among whom all the choices are transitive (Holland & Leinhardt 1977).   (689)

Often terminology that has its source in another is brought into the text, sometimes with or without noting the precise source:

(22)   The "in between" situation that the retiree finds himself in, a "liminal" state, as Turner (1967, pp. 93–111) puts it, is relatively moderate.   (107)

(23)   The ideological formulation of these trends is done by critical groups from the left, those that Cohen (1996) calls "radical liberals".   (586)

The cited others can also be responsible for a theory, approach or model noted in the text:

(24)   Based on the human capital approach (Becker 1975, 1985), which focuses on the individual, and the new structuralism approach (Baron & Bielby 1980; Baron 1994), which ascribes influence on the individual's occupational opportunities to the organization [...].   (637)

They can also be responsible for a finding, which is not worded in terms of an explicit argument, as in the following example:

(25)   A similar trend was found in other countries too (Birkelund 1997).   (654)

The degree of the other's responsibility for the cited content and problems related to the preciseness and authenticity of the other's words will be at the focus of discussion in the following section.

4.2.2 Authenticity and responsibility

In scientific writing, the act of handling other speaker's utterances is less free than in other genres. Writing a scientific text "requires a distinct cognizance of the boundaries of reported speech. It is marked by an acute awareness of the property rights to words and by a fastidiousness in matters of authenticity" (Vološinov 1973:156). For this reason each discipline has a set of conventions for signaling when any alteration, however minor, has been made to the original text (Thompson 1996:512).

We should, however take into account that "the speech of another, once enclosed in context, is – no matter how accurately transmitted – always subject to certain semantic changes. The context embracing another's words is responsible for its dialogizing background, whose influence can be very great" (Bakhtin 1981:240). Moreover, when citing another author, the writer may interpret it based on *his* knowledge, *his* thoughts, *his* theoretic approach, *his* research plan, *his* persuasive goals and so on. In scientific writing, the strategic deployment of the words of others involves what Bakhtin termed 'double-voicing.' In representing the words of others, the author "is simultaneously required to make up a speaking/writing position, to comment on, and evaluate, position themselves in relation to those words" (Baynham 1999:486). These different voices may get mixed up, and the reported utterance may correspond with the original message to varying degrees. As readers of academic papers, we may come across a reference to another author – without quotation marks, sometimes absent a precise reference to a specific page. How much can we infer from it about the cited text? How authentic is the cited context? To what extent can we say that the cited author is responsible for the claim that the present writer has expressed?

These questions can be formulated using Goffman's terms: "animator," "author" and "principal." When looking at the general term "speaker" and trying to define it from the point of view of its social role, Goffman (1981:144–145) differentiates among the various roles included in it:

> **Animator** – the person who is "the talking machine, a body engaged in acoustic activity, […] the individual active in the role of utterance production."
>
> **Author** – the person "who has selected the sentiments that are being expressed and the words in which they are encoded."
>
> **Principal** – the person "whose position is established by the words that are spoken, someone whose beliefs have been told, someone who is committed to what the words say."

When can the author of a paper be considered to be no more than an animator who is merely delivering the authentic words and ideas of the cited author? And when is it clear that the author of the paper is involved? Can we always distinguish between her two roles – that of being the author and that of being the principal? Is she simply the one that has selected the specific words that are being used, or does she have a responsibility for and commitment to their content?

My discussion of these questions will focus on the non-integral pattern, which raises interesting problems of authenticity and responsibility. To what extent has the authenticity of the cited context been preserved? To what extent is intervention of the present writer involved? And as a result – how much responsibility does the writer assign to the cited author?

In Example (26), I have chosen a fairly long text excerpt, in contrast with the simplicity of the invented sentences offered by Swales (1990) and Fløttum et al. (2006) in order to exemplify the different patterns. The complexity of the text, which is a long discourse that makes numerous claims and has a non-integral reference at the end, is one of the main reasons for its ambiguity with regard to authenticity and responsibility.

(26)   Until now, the study focused on meta-cognitive behaviors involved in joint learning. However, it should be assumed that meta-cognitive guidance is effective for the development of problem-solving skills unrelated to the type of learning involved. This is because the meta-cognitive guidance directs the learner to understand the situation of the problem, to note mathematical concepts, to identify the mathematical knowledge needed to solve the problem, to choose the most appropriate strategy, to use control, oversight and reflection for the solution process (Flavell 1979). In fact, meta-cognitive guidance makes it possible to connect two different knowledge structures: quantitative knowledge and situational knowledge related to the degree of familiarity with the situation of the problem. When the two types of knowledge are connected, a mental representation that supports the mathematical model is constructed, and it is that which improves the mathematical performance (Ceci & Roazzi 1994).   (664)

What exactly is Flavell responsible for? Is he responsible for the breakdown of the actions that the learner must perform, listed in the second sentence ("to understand," "to note," "to identify," "to choose," etc.)? Or perhaps the author of the current text is responsible for this breakdown. And perhaps Flavell is also responsible for the claim or assumption that appears in the beginning of the second sentence, that "the meta-cognitive guidance is effective for the development of problem-solving skills." And perhaps also for the claim regarding the causal relationship between the two sentences, expressed in the phrase "This is because..." Or perhaps these claims also belong to the writer of the current text. Is she only the *author*?

A similar situation exists for the Ceci and Roazzi reference: What of all that has been said actually expresses their claims? The division into two types of knowledge? The claim that what makes mathematical performance possible is the mental representation created after the two knowledge structures are joined? All of these points may be claims made by Ceci and Roazzi, and it is reasonable to assume that this is a processed version, a "summary" (Hyland 1999). Moreover, the Hebrew version was written by the current author, and this formulation necessarily contains further processing of the claims. From this it follows that the author cannot, under any circumstances be only the *animator*. She is at the very least the *author*, the one responsible for the wording of the text, and regarding some of the claims, she is also apparently the *principal*.

Thus, the question of authenticity and responsibility remains open. We can speculate, but we cannot be sure. And if we are researchers from the same discipline, with our own scientific agenda and interpretive competence, and we make the effort to look for the cited paper, we may be in for a surprise when we do not necessarily find in it what appears to arise from the quote. Unlike the accuracy and caution that we have come to expect from the scientific discourse, here the reader does not know how to split the responsibility between the current writer and the cited author. This situation is typical of the non-integral citation.

Another problem with the non-integral citation is what Fløttum (2004) called "clusters," namely the accumulation of a number of references in a single sentence. This phenomenon is quite common in academic papers. In the Hebrew corpus 21% of all the citations are of this kind. Here is one example:

(27)   Rejection by peers involves numerous negative features, including isolation, anxiety, sadness, social introversion and depression (Asher, Hymel & Renshaw 1984; Coie, Dodge & Kupersmidt 1990; Vosk, Forehand, Parker & Rickard 1982).   (76)

This text enumerates five features and three references. How do the negative features listed in the text divide up among the various referenced studies?

A further problem is represented by the ambiguous use of quotation marks. Quotation marks have many functions other than indicating a direct quote. They can also communicate various kinds of reservation with the cited utterance, its content or form. (Weizman 1984, 2011). Here is an ambiguous example:

> (28)   The critical discourse in this context emphasizes the manner in which Zionism "invented" the connection to the land and the homeland (Bar-Gal 1993; Ram 1996) and again reveals Judaism's ambivalent approach to place […].   (585)

According to this formulation, we may conclude that Bar-Gal and Ram are writers from the critical discourse studies who are somehow connected to the claim that "Zionism invented the connection to the land." But did any of them actually formulate this claim in this explicit way? Did both of them? Neither of them? It is more than likely that it is the author of this text who is in fact responsible for this formulation, by making a "generalization" (Hyland 1999). Another interesting question is what kind of information the quotation marks around the word "invented" contribute. Do they indicate that one or more of the cited authors used this exact word, or that both of them did? Not necessarily. It is quite possible that the quotation marks indicate the writer's unique usage of the word "invented" (Hebrew: *himtsi'a*), which is in a slightly lower register in Hebrew, and thus creates a controversial claim that is biting and derisive in character.

In this case, we can see that quotation marks have only a limited ability to solve problems of authenticity and responsibility. Examples from the Hebrew corpus demonstrate that the use of quotation marks does not always indicate precisely which part of the utterance can be attributed to another author. Moreover, it cannot even guarantee that such a part even exists.

Even when it is clear that the quotation marks indicate a citation, the problem of attributing the citation to a specific author often persists. For example:

> (29)   This complexity, others note, should be further emphasized on the background of the processes that occurred in Western and Israeli society of the 1990s, which are often described in terms of a "moral crisis," a "value vacuum" or an "age of uncertainty" (Aloni 1998; Sabar Ben-Yehoshua & Gibton 1995; Strike & Soltis 1992; Srarratt 1994).

Here, the writers note three terms in quotation marks, followed by four references. This leaves the reader with no way of knowing how to attribute each of the terms to the various authors.

The use of a non-integral reference in brackets does not always answer the question of which part of the utterance should be attributed to another author either. For example, in (30), there is no reason not to think that the entire claim should be attributed to Gurevitch.

(30)   The development of virtual communities in Israel and the world reflects post-
        modern sociological and cultural processes (Gurevitch 1997).   (314)

The book by Gurevitch is a semi-popular book called "Postmodernism," that doesn't even mention virtual communities. Apparently, it is mentioned here only in terms of general background, in the context of "postmodern sociological and cultural processes." It would appear then in this last example, that in spite of the reference to this author, there is no actual claim in the text that can be attributed to him.

In addition, there are some problems that are unique to the discourse in languages other than English. As I have already mentioned, Israeli researchers in the social sciences writing in Hebrew read most of the relevant literature in English. This is also evident from the bibliographies appended to the papers in the Hebrew corpus: In these lists, only about 25 percent of the items are in Hebrew, with the rest in English. Some writers don't mention even a single bibliographical item in Hebrew, or they mention just one, one that they have themselves written. Thus, the quoted claims in the papers are actually translated, in most cases by the authors themselves, in the course of writing the paper. It is assumed that all translation involves some form of interpretation or processing. This processing can range from minimal (unavoidable) processing as a result of the translation itself, to significant processing that transforms the claims into arguments that may be quite different from the original. We must consider the fact that these researchers are not professional translators, as well as the fact that they are translating the relevant material from their own perspective and for their own rhetorical needs. This further exacerbates the issue of authenticity. I am of course not referring to a deliberate distortion of the arguments, but rather to their perhaps unwitting adaptation in order to integrate them into the new arguments the author is presenting in the paper. In such a case, as noted, the writers in Hebrew cannot be considered animators, but rather only authors, if not principals.

In the soft sciences, especially in qualitative papers, what we see is an effort to develop an idea in context. In many respects, the text of the paper reflects a piece of thought, rather than a piece of reality. Thus, what the citations display for the reader is a network of coordinates, within which the writer seeks to present a new idea. From this perspective, it would not be an overstatement to say that the authenticity of citations is only relative, passing, as they do, through the prism of the writer's rhetorical needs. This last line of thought leads us to the question of the rhetorical or argumentative functions of citations.

4.2.3 The rhetoric of citations

Fløttum (2003) proposes viewing the roles of citations in an academic text on the background of the approach that scientific writing reflects a tension between continuity and progress. This tension between innovation and continuity is connected with the paradoxical tension "between the openness of a scientific text, addressed to the future, and its solid roots in the past" (Beller 1999: 105).

The academic paper looks both backwards and forward: It needs to present itself as a continuation of a particular research tradition, while at the same time offering innovation in the scientific enterprise. The author needs to present new arguments, on the one hand, while showing how these arguments integrate with existing knowledge, on the other. The intensity of this tension varies and is largely dependent on the nature of the innovation that the paper offers. "The more novel the idea, the stronger the need for reinforcement" (Beller 1999: 106). On the background of this tension, citations from the research literature fulfill a number of rhetorical functions.

(a) *Creating a research space.* A large proportion of the citations in academic papers appear as part of the literature review in the first part of the paper (in the introduction, if there is one, or the corresponding section). As many scholars have shown, the literature review helps the author to establish his research space. In the context of doing so, a number of actions are carried out: On the one hand, the researcher describes the existing knowledge on the subject, thereby demonstrating continuity and positioning himself as part of the broader scientific enterprise. Often, he will present a particular approach or theory, describe how others used it and how the current study will continue this research tradition. The continuity can be expressed in the choice of the research subject too (a subject already dealt with by other scholars in the past) or in the choice of the research method (a method that has already serve effectively in the past to generate valuable scientific claims). In this way, the paper turns "backwards," to the past.

On the other hand, the bibliographical references also serve to inform the "niche," that gap in knowledge, that the new article comes to fill. This underscores the progress of the entire scientific project, of which the new claims are part. In this way, the paper turns "forward," to the future.

Moreover, the survey of the research literature is often structured in chronological order. In such a case, the author's perspective, at the current point in time, on the research literature spread out over time is in of itself a reflection of the progress of the research endeavor. This perspective also positions the author at a more advanced point, the only point from which this perspective is possible. In certain disciplines, we can find a type of paper that does not present any new claims at all, but rather only surveys the state-of-the-art in a particular field. This type of paper is attributed with scientific value exactly for this reason: It provides an overview of the progress of the shared scientific enterprise.

(b) *Coherence and innovation.* Another part of the scientific paper in which we are likely to find numerous citations is the last – in the Discussion and Conclusions. At this stage, the main role of the citations is to serve as background for the presentation of new arguments. This presentation can take two directions, and here too the tension between continuity and progress is evident. On the one hand, citations at this stage can help to reinforce the paper's conclusions by demonstrating coherence between the new claims and the previous knowledge (Hunston 1994), for example:

> (31) The findings *are consistent* with surveys that showed that most men discharged from the IDF have a positive view of their service (Meisels 1992a).
> (609)

The cited claims are presented as scientific knowledge that can bolster the speaker's claims, in that he is presenting the development of a consolidated body of knowledge in the creation of which he too plays a role. The coherence between the new and existing knowledge points mainly to continuity, and the innovation requires clarification in relation to the existing knowledge. Nevertheless, it is noteworthy that the factual status of the findings of others mentioned in contexts such as this is not unequivocal.

As I argued in Section 3.3 above, it is not easy to say whether the noting of an authority, namely, attributing the statement to another author, strengthens or weakens the factual status of the utterance. On the other hand, at this stage, citations for the purpose of refutation can also appear, as in the example below:

> (32) The findings of the study *run counter* to the structural theory of role reversal. According to this theory, when the parent is unable to fulfill his roles, the child will take some of those roles upon himself. According to Minuchin (1982), when the parent is perceived as authoritative and responsible, the child will not be drawn into the role of the parent-child. Only when the parental system is weakened does the child take authoritative roles upon himself. In contrast to this, we found that the child may be drawn into the role of the parent-child not only when the parent is weak and not in control.   (435)

> (33) Our findings qualify anthropological impressions from a psychoanalytical perspective, from which it was concluded that serving in combat units contributes to regression to one's youth and impedes the continuation of the maturation process (Yerushalmi 1997).   (612)

In cases such as this, it is the contradiction or qualification that underscore the innovation and progress: The scientific progress is obtained through the conducting of studies whose results either confirm or deny the validity of a particular theory (Example (32)), or refute unsubstantiated claims (Example (33)). In this way, the new research provides more precise and complete scientific knowledge.

(c) *Identity and ethos.* Citations from the research literature help the authors of papers to establish and define their own identity as researchers. By means of the citations, the readers gain a perspective into the research world of the author – to what school of thought he belongs, who he reads and who he holds in high regard. The citations from the research literature help the readers to organize their picture of reality in regard to the relevant discipline, and thereby to situate the author and attach him to a particular group, or alternately, to understand what distinguishes him from other researchers.

Citations are also a valuable means of strengthening the writer's ethos, because through them he can demonstrate his proficiency in the research literature. The author can in this way present his mastery of the literature, which helps him to defend himself against possible counterclaims of ignorance. He can present an ability to judge, criticize and assess other author's arguments, compare different approaches and positions. All these serve as means to strengthen the researcher's ethos to help him obtain the disciplinary community's agreement to the arguments that he is presenting.

4.3 Concession

4.3.1 Introduction

Concession is both a grammatical and argumentative relationship. It is a syntactic and discursive structure, as well as a rhetorical strategy, that has captured the interest of scholars of rhetoric and argumentation since ancient times.

As an adverbial relationship, it is not as frequent as other adverbial relationships such as time, place or manner, perhaps because of its complexity: Its meaning involves other relationships in the background: contrast, cause and condition (Frumuşelu 2007: 425). In Hebrew, as in English, there are two classes of connectives that can express concessive relationships: (a) connectives that specialize in expressing concession, such as *lamrot* (= although); (b) connectives that do not specialize in concession, but which can acquire a concessive function, such as the adversative connective *aval* (= but). The Hebrew *aval*, like the English *but*, the French *mais* and Spanish *pero* can be interpreted as expressing concession in certain contexts (Dascal and Katriel 1977). Thus, concession cannot be identified exclusively by the presence of specialized concessive connectives (Frumușelu 2007: 425).

The complexity of the concession becomes evident when comparing it to adversativity. This comparison will call attention to two fundamental differences: a logo-semantic difference and an argumentative one. From a logo-semantic perspective, adversativity expresses total opposition, whereas concession is characterized by a function of gradual variation. According to Frumușelu, different types of concession are characterized by different levels of adversativity or incompatibility between the content of the two parts of the utterance. What we call adversativity is merely the low point on the incompatibility scale.

From an argumentative perspective, adversativity can be distinguished from concession in that it is a symmetrical relationship, whereas in the concession structure, the two parts of the utterance are not equal in argumentative intensity. I will illustrate this by means of two texts in Hebrew that appeared on the same page in the newspaper, one alongside the other, both relating to the same event: a television report about men who initiated sexual encounters with a 13-year-old girl via the Internet. The report sparked a debate among the public regarding the materials that are appropriate for screening on television news. The two texts were authored by different writers and the purpose of their publication was to present two opposing views on this issue. They were entitled as follows:

(1) Important but contemptible

(2) Disgusting but important (*Haaretz* December 21, 2007)

The article named "Important but contemptible" argued that the screening of the investigative report "points to the fact that the Israeli media have lost their way," and that they should not have broadcast the report. The other article, the one entitled "Disgusting but important," enthusiastically supported the screening of the investigative report, claiming that it was an example of "journalism at its best." This shows that despite the adversative connective, neither of the titles intended to give equal weight to both parts of the utterance. Both utterances, (1) and (2), should be interpreted as expressing a concessionary relationship, which is neither a simple contrast nor symmetrical. The decisive content, that which gives the utterance its argumentative direction, is the content that is introduced by the word *aval*. The value of the concession as an argumentative relationship stems therefore from the fact that one of the parts of the utterance always infers the stronger and final conclusion, and in doing so, overrides the other part, giving the entire utterance a particular argumentative direction.

Following Anscombre (1985), Azar (1997) and Crevels (2000), Frumuşelu proposes four types of concession depending on the type of the negated proposition (Frumuşelu 2007: 431):

> *Content concession* – where the negated proposition is present in the co-text:

(3)   He came and saved my life although he could hardly walk.

> *Epistemic concession* – where the negation affects an entailment that the speaker gets from the co-text:

(4)   He is not at home, although his car is parked in front of his house.

> *Speech-act concession* – where the negation operates on a felicity condition of the speaker's act:

(5)   The answer is on page 200, although I'm sure you already know that.

> *Textual concession* – where the topic of a text sequence is negated:

(6)   A: From which point on did you have the feeling that you'd lost your anonymity?

> B: I still don't consider myself a popular figure, but I think that I have lost my privacy since the time I sometimes hear my name being called from behind me, or catch a look more intensive than other looks… Even though one always gains other things…

A useful framework for discussing concession is Mann and Thompson's Rhetorical Structure Theory, (= RST; Mann and Thompson 1986, 1988) which describes the logo-semantic structure of a text in terms of the functional relationships that exist between two parts of the text, on the syntactic as well as on the discursive level. Mann and Thompson term these two parts *nucleus* and *satellite*: The nucleus is the part that transmits the main content of the utterance, and the satellite is the secondary part. In the case of concession relationships, there is incompatibility between the content of the nucleus and that of the satellite. The concession occurs when the speaker seeks to enhance the positive attitude of the recipient towards the content of the nucleus by means of a satellite, whose content is seemingly inconsistent with the content of the nucleus. The two parts of the sentence are moving in contradictory argumentative directions: The nucleus is the part that is going in the same argumentative direction as the speaker, whereas the satellite is going in the opposite direction.

As an argumentative tool, Azar (1997) distinguished between two types of concessivity that can be considered 'argumentative' in two different ways. His view relies on the concept of 'Topos' (Anscombre 1995), those premises, beliefs or items of knowledge that the participants in a given community share.

(a) *Direct-rejection concessivity* – the opposite of causal relation. If [P → Q] expresses the fact that a causal relation of some kind exists between P and Q based on some topos, then denying one of its portions, i.e. [p → ~ Q] or [~ P → Q] (but not both of them) while maintaining the topos yields a direct-rejection concessivity. For example, on the ground of the topos formulated in (7), one can formulate the causal relations as in (8):

(7)   The colder it is in the room the more people tend to turn on the heat (and the warmer it is, the less they turn it on).

(8)   Since it was very cold in the room, they turned on the heat.

Denying the first portion, the cause, of this causal structure (~ P = 'it was not cold at all in the room') would create the following concessive structure:

(9)   Although it was not cold at all in the room, they turned on the heat.

Denying the second portion, the consequent of this relation (~ Q = 'they did not turn on the heat') would create the following concessive structure:

(10)   Although it was very cold in the room, they did not turn on the heat.

Direct rejection must by definition contain a proposition contradicting the cause (P) or the consequent (Q) of an expected causal relation. This kind of structure can be used for both argumentative and nonargumentative purposes according to the context, but even in argumentative contexts, it can be considered 'argumentative' only in a weak sense. Following Chittleborough and Newman (1993), Azar called this kind of concession 'persuader'. A 'persuader' is a "psychologically manipulative technique used by an arguer with the intention or hope of increasing the conclusion being accepted by a recipient" (Chittleborough and Newman 1993: 196).

(b) *Indirect-rejection concessivity* – the two portions of the text express two different arguments leading to two opposite conclusions, which are not explicitly stated and must be inferred. For example:

(11)   True, his car has broken down, but there is always public transportation.

The first portion of the text may lead to the unexpressed conclusion formulated in (13), on the ground of the topos that is formulated in (12):

(12)   People with faulty cars may encounter difficulties in getting to work.

(13)   He could not go to work.

The second portion of the text may lead to the unexpressed conclusion in (15), which is based on the topos formulated in (14):

(14)   The more public transportation is available, the more easily people can go wherever they wish.

(15)   He could go to work.

Note that this indirect-rejection concessivity can be changed into direct-rejection concessivity by explicitly stating what is only inferred in it, for example:

(16)   True, his car has broken down, but since there is always public transportation, he could have gone to work (Azar 1997: 312).

The two adversative arguments, based on two different topoï, are always 'argumentative' in the stronger sense, i.e. reason-type arguments. Azar called this type "supportive argument," while being 'supportive' means "being a reason or item of information presented in an argument which is intended to provide support for a conclusion" (Chittleborough and Newman 1993: 198).

In another place, Azar (1999) distinguished between the argumentative and the circumstantial concessions. This distinction is based on how the Rhetorical Structure Theory (Mann and Thompson 1986, 1988), defines the concessionary relationship as relationship in which the speaker seeks to enhance the positive attitude of the recipient towards the content of the nucleus by means of a satellite, whose content is seemingly inconsistent with the content of the nucleus. According to Azar, unlike in this definition, the concession structure is sometimes used only to describe that state of affairs in the extra-linguistic world, and in such a case, it does not seek to persuade, since "there is no point in talking about persuasion when simple facts that are known to everyone or will soon become known to those that are interested" (Azar 1999: 290). In such a case, the satellite presents a framework in light of which the reader interprets the state of affairs presented in the nucleus. A satellite of this kind would not be considered a concessionary satellite, but rather a circumstantial one, which represents a different type of logical-functional relationship, as described by the Rhetorical Structure Theory.

This is incompatible with the general approach of the current study. As I noted above in Section 3.3, an academic paper does not generally include "simple facts that are known to everyone." The factual status of the utterances in this genre is far more complex. Moreover, having defined the academic paper as a persuasive text (in Chapter 3 above), I perceive each of its parts as a means that has the potential to promote its persuasive goal. Take for example the sentence:

(17)   Despite the central position the subject occupies on the public agenda, it has yet to be investigated in depth.

According to Azar's distinction, this sentence could serve as an example of a circumstantial concession rather than a persuasive one, since it seemingly involves the transmission of two facts, and the fact transmitted in the first part is presented as the circumstances in the context of which the transmitted content of the second part should be understood. However, this description appears to me to be incomplete. First, regarding the matter of "facticity": Note that both parts contain an element of a relative nature: The "centrality" of a subject, even if it is consensual, is of a relative nature, and the same goes for the degree of the study's "depth." Consequently, while these two propositions may be viewed as "facts" in the context of a day-to-day conversation, it is clear that in the context of a scientific argument, this is a formulation that has been deliberately and cautiously selected in order to serve a goal of some kind. Second, as for the goal: It is easy to see how an utterance such as this serves a clear persuasive goal in the scientific discourse – the design of the research space. Thus, it would appear that the concession that Azar calls "circumstantial" is merely direct-rejection concessivity, which can be considered to be on the lower end of the argumentative scale, i.e. it can serve as a rhetorical device to advance the author's aim. These goals will be presented with examples in Section 4.3.2.

345 examples of the concessive structure were found in the corpus, and they include three of the four types proposed by Frumuşelu. Here are a number of examples:

(a) *Content concession.* As noted, this type of concession requires the negated proposition to appear explicitly in the text. Such an example can be found in the following sentence:

(18)   Despite the fact that the doctors and educators in Eretz Israel of the early 20th century held different views, and although the discourse was distributed among the different social sectors in Eretz Israel (parents and children, boys and girls), and despite the use of different strategies such as lessons, treatments, lectures, pamphlets and books – considerable similarity could be found in the textual content dealing with sex education at that time.   (539)

In order to analyze this structure, it could be given a simpler form:

(19)   People holding different views, coming from different sectors, using different strategies created similar textual content.

This content is interrelated to an expected causal relationship that could be formulated as follows:

(20) People holding different views, coming from different sectors, using different strategies can be expected to create different textual content.

Sentence (19) is in fact a negation of the expected result:

(21) People holding different views, coming from different sectors, using different strategies created similar (=~different) content.

The concession structure then includes a rejection of the second portion of the topos. In this, the structure resembles Azar's direct concession. Azar maintained that this type cannot be considered argumentative in its strongest sense, and it indeed appears that in this case, there is no argumentation behind the claim. However, this example can be viewed as argumentative in the weaker sense. The claim, that some similarity was found despite the great difference, supports the paper's chief claim in general terms, as well as its research method.

(b) *Speech-act concession.* Because the scientific text is made up (almost entirely) of assertions,[4] it is only natural that it is difficult to find examples of other speech acts. I will propose then just one example, which includes reservations with a representative speech act (= assertion). The assertion expresses a conclusion, and the reservation that is expressed in the concession is reservation with the conclusion.

(22) Despite the fact that from the current analysis, it is not possible to draw unequivocal conclusions regarding the causality, it would appear that mastery of the language is a relatively important integration channel [...].   (140)

We can see the reservation with the speech act if we surface the explicit operative verb:

(23) Despite the fact that it is not possible to draw unequivocal conclusions, [I conclude that] it would appear that mastery of the language [...].

However, the vast majority of concessional structures are of the type that Frumuşelu calls epistemic concessions, the type that corresponds with Azar's indirect-rejection concessivity, which is argumentative in nature. These structures are always connected to the author's position and his effort to direct the discourse in the argumentative direction she wants it to take. The substantial distribution of concession structures in academic papers provides additional indication of its argumentative nature. The following section will be devoted to a further elaboration of this subject.

---

4. I will elaborate further on speech acts of questions in Section 4.5.

4.3.2 The rhetoric of concession

In this section, I will describe the concession structures found in the corpus on the background of the academic paper's persuasive aims as presented in Chapter 3.

(a) *Creating a research space.* In general terms, it would appear that the concession, as a structure having two parts that pull in opposite directions, may reflect the tension that exists between the writer's two opposing aims. As such, it is particularly suitable for the design of a research space. When creating a research space, the author should mark out territory within which the study will be carried out, with a considerable emphasis given to the centrality of this territory, and a description of the niche in which the current study will be able to present its new argument. However, in order to establish the significance of the actual research, the author should argue that this niche is neither too minor nor marginal. These claims, in favor of both the importance of the subject and the existence of a niche, are somewhat competitive: In order to persuade the reader as to the centrality of the subject, it is necessary to survey what has already been written about it. However, a survey of the relevant literature may give the impression that the subject has already been exhausted and that there is nothing to be gained by any further research. On the other hand, if attention is focused on what was ignored by other scholars, it could create the impression that it is simply unimportant. This built-in tension between these two goals invites special persuasive effort in two different directions, which is given natural expression in concession structures.

It should consequently come as no surprise that Swales (1990: 154) found that in a large number of papers, the design of a research space invites the appearance of adversatives such as: *however, nevertheless, yet, unfortunately* and *but.* Many of the concession structures that I will present below create a contrast between the two aims in a way that in general terms and in a somewhat simplistic fashion can be described as follows:

(24)   Although the subject has been the target of research in general, there still remain aspects that have not yet been explored, or a new perspective from which it can be explored.

If we add to this description the implicit conclusion that arises from the satellite part and is rejected in the nucleus, the following complete picture will be received:

(25)   The subject has already been the target of research [which is why new research is not important], but there still remain certain aspects that have not yet been explored [which is why the new research is important].

This then is a clearly argumentative textual format through which the importance of the research can be justified and highlighted. We can see this in the following example:

(26) *Although* the last wave of immigration from the CIS has been the subject of intensive research (Leshem & Shor 1997), only a few individual studies have examined the adjustment of teenage immigrants from the CIS in dormitories. This study will report on the connection between cultural identity and psychological and social adjustment in the special context of dormitory life.
(200)

The research that has been carried out until now is presented in the satellite position in general terms: "has been the subject of intensive research." On the other hand, the niche, in which the current study will present new arguments, is designed by means of specific components that are included in the nucleus: "teenage" and "dormitory life."

The niche within which the study will make new claims can be described in terms of observation of a similar object of research, but from a different perspective and by raising new research questions, which were not asked in previous studies. Let's see a number of examples of this:

(27) On this background, it can be understood why the leaving rate served – and still serves – as the central index for the measure of solidarity, the state's attractiveness and social empowerment (Cohen 1999) […]. This study also places the "test of leaving" at the center, *but from a critical perspective.* (573)

(28) In this study, we once again explored the impact of family size on the number of years of schooling, *but for the first time in Israel*, we also examined its effect on IQ, scholastic achievements and stratified ambitions.   (195)

(29) In England, Cooper and Kelly […] discovered five factors […]. *Nevertheless*, no attempt was made in their study to look for the connection between these stress factors and manager burnout.   (222)

(30) The feminist aspect of the advertisements screened on this channel has been placed on the public agenda too. *However*, the prevalent treatment for the subject was localized, i.e. it focused on specific advertising announcements.
(470)

The existence of a niche that has not yet been explored through research is not enough to make its study valuable. A further potential claim that the researcher must contend with is that the niche has not been studied because it is too insignificant. The following example shows one way of contenting with this implicit claim:

(31)   The article focuses on a short episode in the history of sex education, a story
that has not yet been told. *Nevertheless*, it represents more than a mere curiosity; it is a link in a dynamic continuum in the history of sex education [...] in
Israel.   (533)

The fact that the story "has not yet been told" and that it is a "short episode", may
lead one to conclude that this niche is extremely minor and is therefore one that
is not worth researching. This argument, if raised, poses a real threat to it being
accepted as a valuable piece of research. In order to refute this claim, the author
mobilizes the concession structure – it presents the claim in the satellite and rejects it in the nucleus by saying: "It is more than a mere curiosity (= Hebrew:
mutzag kuriozi); it is a link in a dynamic continuum."

Yet another potential claim is that the general subject is not important or
relevant enough. This claim, should it be accepted, can diminish from the study's
importance, and numerous concession structures are mobilized to contend with
it. Let's look at an example:

(32)   *Despite* evidence from the research regarding the decline of the older parties
in Israel (e.g. Goldberg 1988), it is interesting to note the ethnic contexts of
party identification.   (19)

The implicit conclusion that is presented in the satellite can be understood as
related to the importance of the study: If the older parties are declining, perhaps
the question of party identification is no longer relevant or interesting. If the current study deals with this question, it is then an irrelevant and uninteresting one.
The role of the nucleus is then to reject the implicit conclusion that arises from
the satellite. The schematic structure of the concession as an argumentative device
can be described as follows:

(33)   The older parties are declining [which is why a study on party identification
is not relevant], but it would be interesting to examine the ethnic contexts of
the party identification.

This schematic description shows that here the claim that is supposed to reject the
conclusion implicit in the satellite ('there is an aspect that may appear to be irrelevant') is not a conclusion that arises from the nucleus, but is rather a claim that
the content in the nucleus explicitly argues with by means of the evaluative adjective 'interesting' (= Hebrew: *me'anyen*). This type of use of evaluative adjectives to
emphasize the importance of the research can also be seen in Example (34):

(34)   The *fascinating* encounter between army service, as an experience that has a
powerful effect on the lives of people (Elder 1985), and youth as a chapter in
the maturation process, has not yet been the subject of a study as it deserves.

> In Israel too, despite the fact that service in the Israel Defense Forces is so *central* to the lives of so many young people, very few studies have explored it as a formative chapter of life affecting personal development.   (594)

Here, rather than establish the importance of the subject by surveying the relevant literature, the authors use the adjectives 'fascinating' (= *meratek*), 'powerful' (= *chazaka*) and 'central' (= *merkazi*), which present the importance of the subject as a fact that contrasts with the fact that it has been the subject of little research.

The concession structure in the second sentence appears to be an example of a direct concession. The content in the nucleus does not run counter to the conclusion that arises from the satellite, but rather denies the causal relationship that could be described by means of the topos:

(35)   The more interesting, central and significant a particular phenomenon is, the more we can expect it to be studied.

The fact that the phenomenon under discussion has been the target of little research is then just the opposite of what one might expect. In the case of a direct concession, the claim can be defined as argumentative only in the weakest sense. There is no real argumentation here, but rather the creation of an impression of importance in order to justify the research. The importance of the subject is presented by means of evaluative adjectives, and the fact that it is nonetheless being studied is presented as seemingly contrary to logic.

The importance of the study can be based on an opposite direction too: Instead of arguing that the subject has been explored in general terms, but that there remains an available niche, it can be claimed that the previous studies together filled a number of individual niches, but that no complete picture yet exists. Here is an example of the creation of a research space in this way:

(36)   An examination of the professional literature shows that *there is no one single comprehensive and thorough* study on grief in Israeli society. *While much has been written* on certain aspects of grief, including on commemoration and memorial stones (Shamir 1996), the poetry of the War of Independence (Meiron 1992), the care for the bereaved (Malkinson Rubin & Witztum 1993), the eulogy (Kasher 1993) and others, and *despite* the valuable contribution made by these studies to the understanding of *part* of the mosaic of grief in Israeli society, none has *yet* presented a *complete* picture of the mosaic.
(281)

By using the concession structure, the authors present a contrast between the partial picture presented by the existing studies and the complete picture that they plan to present in the current one. The importance of the study is emphasized here by means of the complete picture that it seeks to present.

(b) *Justifying the choice of the theoretical framework*. In the social sciences and the humanities, it is not unusual for competing theories to coexist alongside one another. Many scholars choose a theoretical framework for the purpose of a specific study, in accordance with the subject of the study. This choice requires an argumentative effort to justify it, and the justification may be presented in a concession structure, as in the following example:

(37)   *While* the structural theory explains the circumstances in which one of the children in a family takes a parental role upon himself, it does not make clear what the effects of this role are. [...] In order to better understand the personality and emotional structure of the parent-child, we have used the attachment theory, which also relates to the phenomenon of the parent-child.
(416)

The use of the concession structure enables rejection of the theory that was not chosen. Within the concession structure itself, the advantages of that theory are brought to the surface (by saying that it can provide an explanation). However, its disadvantages are mentioned immediately afterwards. The disadvantages receive priority because they are presented in the nucleus, and the result is a rejection of this theoretical framework as unsuitable. Further on, the reader is told about the other theoretical framework, the one that was chosen. The schematic structure of the argument could be formulated as follows:

(38)   The structural theory explains the circumstances in which one of the children in a family takes upon himself a parental role [and thus this theory can fit our research], but it does not make clear what the effects of this role are [and thus it does not fit our research].

In contrast, in Example (39), the concession structure serves not to reject a competing theoretical framework, but rather to reject possible claims against the approach that was in fact selected:

(39)   *Even if* the conditions that Jameson offers still need to be bolstered empirically, and *even if* the connection between the psychological facts that he notes for collective activity are not necessarily a causal relationship, their very existence is important, if only to address the existence or nonexistence of a social rift.
(10)

The following example illustrates the status of the definition as part of the theoretical framework. In case there are a number of definitions for the same matter, the choice of the right definition for the purpose of a specific study often requires justification too. The concession structure is appropriate for the logical structure of rejecting a definition while weighing its advantages and disadvantages. The disadvantages will be highlighted and take on a preferred status if positioned in the nucleus, as in the following example:

(40) A definition of this kind offers a framework and direction that should be chosen when seeking to identify groups within a social unit; *however*, it does not yet define an unambiguous process that enables the identification of groups.

(687)

Presenting the disadvantages of the definition in the position of the nucleus leads to its rejection, which enables the acceptance of a different definition.

(c) *Justifying the methodology.* An examination of the concession structure in the Hebrew corpus showed that many of them are related to questions of methodology. Let's look at the following example:

(41) This finding, *despite* the relatively small sample (which invites further confirmation), supports the interpretation that a real change in the perceptions of the immigrants has occurred [...].   (142)

This utterance refers to a study based on surveys. The phrase "despite the relatively small sample" contains an explicit qualification regarding the size of the sample chosen for the surveys. The claim that the sample is small (or too small) may pose a genuine threat to the reliability of the study's findings and their significance. The authors could have ignored this point, but instead, they chose to bring it to the surface and insert it into a concession structure, in a satellite position. By doing so, they manage, on the one hand, to show that they are aware of the problem but, on the other, to reject the claim and not allow it to weaken their position.

From the wording of Example (41), it is clear that the researchers themselves accept the claim that the sample that they selected was small. This claim is in fact based on the agreements shared by the members of their disciplinary community, for example agreements regarding how big or how comprehensive a sample should be in order for the conclusions to be considered valid. Thus, the reservation with the size of the sample fulfills a dialogic function in reconfirming the disciplinary agreements. The acceptance of this claim by the researchers underscores their membership in the scientific community to which the potential readership of the paper also belongs, and reconfirms the agreements they all share. The utterance in parenthesis – "which requires further confirmation" – also serves a dialogic role not only in that it raises a claim that any of the other members of the discourse community could raise, but also in that it is perceived as inviting all of them to join in the research endeavor and examine the question themselves.

As in this last example, many papers explicitly note the weaknesses of the study's methodologies. These explicit statements can be transmitted in the concession satellite, as in the three following Examples (42)–(44). The content in the nucleus position shows that the researchers are aware of these disadvantages and have resolved them in one way or another.

(42) Methodologically speaking, *it would be better* if we could relate to all the elements of the same individual profile. *But* because we do not have data from each interviewee on all the details that were explored, we had no choice but to compose a type profile from various interviewees, and this should be related to as the ideal, Weberian type.   (116)

(43) This possibility can be viewed as an indication of a *certain weakness* of the super-analysis method applied here, in the context of conditioned variables. *In practice*, we are concentrating on the cases in which an analysis according to a conditioned variable diminishes the variability of the correlations, and relates to the other possibilities […] as the absence of any indication of the effect of this variable.   (62)

(44) In the current survey, *data were not collected* that would enable an examination of the extent of the appearance of changes in the way Israeli society relates directly to bereavement, *but*, when analyzing the results, we ask ourselves if the significant changes in this attitude are a natural phenomenon.   (282)

Noting the methodological weaknesses of the research plays an important role in the dialogue with the discourse community. Their appearance in the text can, on the one hand, present the researcher as thoughtful and cautious, as one who has himself considered the possible reservations that might be raised regarding his arguments, and addressed them; on the other hand, they are likely to contribute to the solidarity of the community through the confirmation of the agreements and assumptions accepted by its members.

Each methodology has advantages and disadvantages, and in many cases, researchers find it necessary to justify the choice of one methodology over another. In Example (45), the concession structure, in the satellite position, raises possible reservations related to the study's methodology. The author here makes it clear that he is aware of the problems posed by the method he has chosen, although he does not point out what those problems are. The concession structure helps him to raise the reservations in a laconic fashion in the satellite and immediately reject them in the nucleus, and thus avoid weakening his argument:

(45) *Despite methodological problems* involved in the use of questionnaires to measure variables related to the intensity of ethnic awareness and identity, the answers to identical survey questions […] can help to identify trends in the ethnic rift of the past decade.   (6)

In contrast, a concession structure can serve to reject the research tools used in previous studies, which the author decided not to use in the current study. We will now look at a number of excerpts from a paper describing a study on the effect of domestic physical violence on the adjustment of children in school. In order to investigate how children function socially in school, the researchers used two data sources: peer assessment and teacher assessment. These are not the only possible sources, and consequently their choice needs to be justified. The following excerpt appears in the first part of the paper, in the section that surveys the possible sources of knowledge used in various studies in the past. If we follow the concession structures, we will note an effort to reject the inclusion of the data sources that were not chosen for the current study.

(46) *Children*: Self-reports can produce knowledge that is not known to others and that precisely reflects the child's values […] *but* which do not show us how others perceive the child's adjustment.
Parents: […] Parents provide rich data on the behaviors that occur in the home. *However*, because so much of the child's day is spent in school and in other activities outside the home, essential events that can shed light on the child's status could be unknown to the parents […].
Observers: […] The unique aspect of observation lies in the ability to distinguish between nuances, such as the extent to which children play on their own […]. *On the other hand*, observations of children's behavior could produce misleading data.   (77)

Immediately following, the data sources that were chosen, friends and teachers, are noted; however, various methods for data collection are mentioned, with those not chosen for the current study being rejected:

(47) *Friends*: Price and Van Slyke (1991) asked children to rate the extent to which they liked each child in their group. This simple method is sufficient when working with young children, *but is limited* because it estimates only the emotional assessment of the participants. […] Dodge et al. (1990) studied the perceptions of friends […] by asking the children to name up to three children that fit each of three behavioral descriptions. […] This information may prove to be very effective, *but it provides a very narrow* understanding of the social capabilities of the child that is at the focus.

> *Teachers*: [...] Reports of this kind make it possible to assess the children's achievements, advancement and behavior in school. *However*, in most cases, the information is not based on standardized questions that enable a comparison with friends and could therefore reflect teachers' *biased* assessments.
>
> (78)

The authors then mention a number of possible research tools that could be used to investigate their research question. They explicate the advantages of each in the satellite, but immediately go on to enumerate the disadvantages in the nucleus. The adjectives 'limited' (= *mugbelet*), 'very narrow' (= *meod tzara*) and 'biased' (= *mutot*) with their negative semantic value indicate the weakness of the methods that are mentioned. The result is a rejection of these methods, which is followed by a presentation of the method that they opted for.

The concession structure serves then as a major instrument among writers to create a dialogue with previous studies and justify the choice of a particular research method. This justification is of supreme importance when bolstering the validity of the claim. The concession structures lead the reader through the following argumentative path:

> Mention of further sources of data > their rejection > mention of data sources that were chosen with the mention of other research methods > the rejection of the other research methods.

This argumentative path implies the superiority of the current study. The detailed discussion of concession structures used in this paper demonstrates the important role of the concession structure in justifying the choice of the research method, while holding a dialogue with previous studies and accepted research methods.

(d) *Justifying the statistical analysis.* As noted earlier, Hunston, in her analysis of the argumentative aims of the various parts of the academic paper, identified the goal of the Results section as being to prove that the statistical analysis was adequate and appropriate. The presentation of quantitative findings in a paper is always accompanied by a verbal explanation of those findings. The figures do not "speak for themselves," and the researcher must "speak" on their behalf. In this context, she must choose which data to emphasize, and how to present them in order to give them meaning (Perelman and Olbrechts-Tyteca 1958: Section 30). Let's look, for example, at the following sentence, excerpted from a paper on the representation of women and men in television commercials:

(48) The use of voyeurism [...] is also seven times more frequent for women than for men, *although* the overall rate of such scenes in absolute terms is low.

(479)

The researcher notes here two findings that arise from the number tables he has presented:

(49)   The use of voyeurism is seven times more frequent for women than for men.

(50)   The overall rate of such scenes in absolute terms is low.

The concession presented here could be considered an example of a circumstantial concession rather than a persuasive one: These are two facts presented one alongside the other. Moreover, one might even claim that there is no concession structure here at all, but rather a contrast structure, notwithstanding that the connective chosen here ("although" = *af ki*) is a typical concession connective. However, based on the assumption that every linguistic choice can have an argumentative value (even if in a weak sense), I would suggest that even in this case, it is relevant to talk about the argumentative use of the concession structure. The formulation that the author chose positions the first finding in the nucleus, and the second in the satellite. Because the nucleus position enjoys a preferred status, the first finding will be given greater weight. This is evident if we switch the positions of the findings within the concession structure:

(51)   *Although* the use of voyeurism is seven times more frequent from women than for men, the overall rate of such scenes in absolute terms is low.

This wording, which presents not the difference in frequency but rather the fact that the absolute numbers are low in the preferred nucleus position, creates a slightly different reading of the exact same data. From this it follows that the verbal description not only lends meaning to the data, but that the choice of a particular wording is a conscious one that makes it possible to use the data argumentatively, i.e. to give them an argumentative direction that supports the researcher's claims and allows her to move ahead in presenting her argument. Among other things, this choice can highlight particular findings so as to demonstrate that the researcher's choice of a certain type of statistical analysis was justified and led to valuable findings. In the three examples below, all of which contain a typical concession connective, we can see how the researcher places in the nucleus position the finding that demonstrates that the statistical tests "show" something that can be considered valuable – that it points to the existence of a particular phenomenon (Example (52)), dynamism (i.e. the identification of the existence of changes or processes – Example (53)), or the identification of connections between different phenomena (Example (54)):

(52)   The first component, identity, exists among most people of Mizrahi extraction (by virtue of their self-identification as being of Mizrahi extraction rather

than absence of an ethnic awareness), *even if* only in a minor form (which is expressed in the low identity rating it is assigned in comparison to other identities).   (16)

(53) The diagram shows that the past ten years has seen a growing sense among people of Mizrahi extraction (according to their own definition) of the image of the parties as narrowing the ethnic divide, *although* in 1996, the Likud was still clearly perceived as the leading party where narrowing the ethnic gap was involved.   (20)

(54) The upper part of Table 2, which relates to the entire sample, shows that the way the immigrants perceive their own mastery of Hebrew is related to the first date for most of the other indices of assimilation that were examined, *although* the connections are not strong. The connection between their mastery of Hebrew and their social involvement and job status is Israel existed after a year too, *although* it was weakened.   (140)

In all these cases, a further finding is placed in the satellite position, which is also significant. Its positioning in this place does not detract from its status as describing the "facts," although, as noted, it weakens its argumentative force.

Another example is taken from the Conclusions section of a quantitative paper. However, the concession structure in it supports not only the justification of the conclusions and highlighting of the innovation in the paper, but also works to justify the statistical approach that was chosen and proves that it can arrive at valuable conclusions. It implies that the research tools and the statistical analyses carried out were chosen cautiously:

(55) Among the explanations for inequality between the sexes are claims that the disparities are partly the result of different human capital and discriminatory practices in the work market. The background characteristics of the men and women in the present study are quite similar both demographically and academically. *Despite this*, we found already at the start of their careers, that the proportion of men filling senior positions was higher than the proportion of women. Consequently, the differences that were found are indicative of discriminatory treatment in the work market, i.e., it is possible that men and women having similar human capital will have different chances of advancing to senior positions at the start of their professional careers.   (277)

The first sentence begins by noting two claims that appear in the professional literature. The findings of the current study tend to refute the first claim. The refutation is transmitted in a concession structure, which has the ability to emphasize the cautious wording of the research question and nature of the statistical approach that was chosen: The claim that is in the satellite position ("The background characteristics of the men and women in this study are quite similar") together with the claim quoted from the research literature ("the disparities are partly the result of different human capital") raise implicit conclusions ("One would not expect to find differences between men and women"), a conclusion that is immediately rejected in the nucleus position. The schematic structure of the concession can be expressed as follows:

(56)   The disparities between men and women are the result of different human capital and in the current study the human capital of the men and women was similar [and consequently, one would not expect to find differences between the sexes at the beginning of their careers]; *however*, differences were found between the sexes at the beginning of their careers.

The fact that the findings of the study are able to reject the expected result underscores the innovation that they offer and is indicative that the statistical approach chosen for the study was successful.

(e) *Justifying the findings' interpretation.* In quantitative studies, the findings themselves are often the result of interpretation. We will see this in the example below. In the findings section, the researchers provide a detailed description of the external appearance of a certificate the IDF awards to those leaving the army – the type of paper, the font used, its color, etc. – and propose a symbolic meaning for each of these details. The description concludes with the following words:

(57)   The full range of these symbols closely associates the IDF and its wars with traditional elements involving a link to the distant past (parchment) and more recent (IDF wars) history. This interpretation is not the IDF's official interpretation, *but* it appears to us that the designers of the symbols were working in accordance with these codes, even if they did so unconsciously.   (109)

Here the argumentative effort to justify the proposed interpretation is straightforward. The concession structure places a reservation that could come up in relation to this interpretation in the satellite position, thus weakening its strength. (The researchers would certainly prefer to find support, even indirect support for their view that the certificate is designed in accordance with explicit instructions that would prove that the IDF views these details as having similar symbolic meaning.) A further reservation is placed in the nucleus position by means of the hedging phrase "it appears to us." Especially notable is the phrase "even if they did so unconsciously," which has considerable argumentative value: The claim that something was done based on an unconscious intent is a claim that although difficult to prove is no less difficult to refute.

(f) *Justifying the conclusions.* The next stage in the claim is drawing conclusions from the findings and interpreting those conclusions. Drawing conclusions is a "leap" into a new area in which the "objective" ground is far less solid. There is good reason why we find numerous hedging expressions and those conveying caution in the Conclusions section (Lewin et al. 2001). The greater the leap, the greater the argumentative effort that is needed, in addition to more cautionary measures.

In the following example, the conclusion is based on a quantitative finding, which expressly supports it, and consequently the leap is not large. The conclusion first appears in the sentence before the concession structure, worded so as to reflect how "natural" and independent of the researcher it is ("The results of the study point to…" see Livnat 2010a). Then, within the concession structure, it appears once again in the nucleus position together with the quantitative finding that supports it. In the satellite position are further quantitative findings that might ostensibly weaken the strength of the conclusion. A number of components are responsible for the fact that this potential weakening is not realized, but is instead immediately rejected: Besides the fact that the conclusion itself is based on a quantitative finding, it is important that the seeming reservations appear after the conclusion has already been explicitly formulated, embedded between two different formulations of it. There is also some importance to the appearance of the predictive concession word "although" before each of the reservations, because it enables the reader to anticipate the rejection that will appear immediately after:

(58) The findings of this study clearly point to the fact that aggression was not mediated by any internal state of the subject. *Although* it was found that the personal and public sense of awareness among the individual subjects was higher than for many groups, and *although* the results of the study point to a higher level of aggression among the many as opposed to the individuals, it would appear that this higher level of aggression is not the result of a sense of awareness, because in the covariance analysis, no connection was found between the internal factors and the aggression level. Moreover, the findings of the study point to the influence of the "anonymous" factor and the "relationship" factor on the type of identity of the victim.   (409)

Contrary to this case, in other cases the conclusion does not emanate from the quantitative findings; On the contrary, the researcher seeks to draw the conclusions despite a deficiency of sufficient quantitative findings to support them. In this case, the leap is greater and consequently, the conclusion will be worded more cautiously. In the following example, the adverb 'relatively' (= *yachasit*) is notable, along with the hedge 'it appears that' (= *nir'e ki*). The concession structure also plays a clearly defined argumentative role here. The conclusion will be positioned in the nucleus, whereas the reservation will be placed in the satellite position so as to weaken it:

(59)   *Although* causality cannot be *unequivocally* concluded from the current analysis, it appears that mastery of the language is a *relatively* good channel for the immigrants' assimilation into Israeli society [...].   (140)

The reservation raised in the satellite is weakened by means of the adverb 'unequivocally' (= *chad-mashma'it*) which implies that even if the findings do not support the conclusion, they do not rule it out either. The adverb "directly" has a similar role, as can be seen in the following two examples (both taken from the same paper):

(60)   *Even if* it is difficult to infer *directly* from these tests regarding the possibility of collective activity or uprising on the background of the ethnic divide, a great deal can still be learned from them about the political significance of ethnicity in Israel in the last decade of the 20th century.   (19)

(61)   If so, we cannot infer *directly* from these data regarding an anticipated increase or decrease in ethnic tensions in Israel, *although* it appears that they are indicative primarily of the continued existence of the ethnic divide.
                    (24 25)

The word 'directly' (= *yeshirot*) implies that it is possible to draw the desired conclusion indirectly, and consequently weakens the reservation that appears in the satellite. In the following example, the reservation is weakened by the hedge 'may.'

(62)   *Although* multiple identities *may* contain considerable potential for conflict (for example, between the cultural perceptions related to one's Israeli identity and the original one), it appears that the ability to maintain a number of identities, despite the contradiction, enables the immigrant to expand his circle of social, value-based and even material resources, which can contribute to his wellbeing.   (214)

(g)   *Presenting an innovation*. Towards the end of the paper, a persuasive effort related to the need to present new claims is evident, after they have been presented in their entirety as valuable claims that contain an innovation. The concession structure can emphasize the innovation as being related to the rejection of a particular theory, because the findings are incompatible with the theory's projections, for example:

(63)   Theoretically, according to the human capital theory, academic achievements can contribute to occupational opportunities at the beginning of one's career. *However*, we found no support for this claim in our study.   (276)

The concession structure can also emphasize the innovation by pointing out that a particular phenomenon is more important than it is conventionally thought to be:

(64) *While* the "resource availability" pressure factor can be found in the literature [...], it is mentioned in only a few isolated studies.   (239)

In some cases, the concession structure appears following an entire section of text in which possible reservations with different parts of the claim are raised. The following example is taken from the final paragraph of a paper, the paragraph that culminates the Conclusions section. In this case, the concluding paragraph includes an explicit mention of the reservations, in particular those that are the result of the limitations of the methodology. The mention of the reservations ends in a concession structure that places them in the satellite position ("Despite these possible reservations"). The innovation that the research proposes is presented in the nucleus position, whose generalized wording is aimed at overriding all the reservations:

(65) *Despite* these possible reservations, this survey should be viewed as an *initial* effort to explore important issues that are not sufficiently well known [...].
    (296)

It is interesting to note that the nucleus is worded in accordance with the niche described at the beginning of the paper: The phrase "important issues that are not sufficiently well known" is related to the creation of the research space – the importance of the subject and the existence of the niche. The adjective "initial" (= *rishoni*) is especially interesting: On the one hand, it marks out the importance of the study, but on the other, expresses the possibility of incompleteness, because something that is "initial" could also be premature or of a probing nature, whose importance lies less in its completeness than in the fact that it is the first. This is also hinted at in the next sentence, which indicates that the niche as a whole is still open.

All the objectives presented in this section are clearly argumentative ones and are indicative of the effort made by the researcher to present the new knowledge as deserving of being accepted as part of the shared disciplinary knowledge base. The concession structure is then an important argumentative structure, which supports a wide range of persuasive goals that are typical of the academic paper.

4.3.3 Concession as dialogue

Concession is one way of implicitly acknowledging the presence of the others. Fløttum et al. (2006: 242) distinguish between "explicit polyphony" through bibliographical references and "implicit polyphony" that may be expressed in the text, among other constructions, by concession. Concession (both as a syntactic structure and a discourse structure) is one of the ways in which counterclaims are presented and rejected. This includes objections that are actually voiced, those that echo the author's inner reasoning and deliberations, as well as those that the author presumes are likely to come up following the publication of the text. In some cases, the counterclaim is not attributed to a specific individual, but rather presented in general terms as an argument that any reader could raise to counter the claims made by the author.

The dialogicity through concession may well be explicit in case of an explicit counterclaim. Let us look at an example from the Hebrew corpus in which the rejected claim is raised and explicitly refuted. The example is excerpted from a study that deals with the ability to predict success in academic education. In this study, the criterion chosen to represent academic success is the participant's first-year university grades. The following paragraph attempts to justify the choice of first-year university grades as a criterion of success.

(66) It is worth relating parenthetically to the oft-heard claim that the most suitable criterion for validation is the average grade at the completion of studies towards the undergraduate degree rather than the average grade at the end of the first year. [...] *Even if* there is some justification for the position that the average grade at the completion of studies is a more relevant criterion for validation than the first-year grades, there is no empirical basis for the claim that often accompanies this position, that the predictive validation for this criterion is lower than the predictive validation for first-year grades [...]. These findings, in addition to the arguments noted here above regarding the advantages of employing the first-year grades as a criterion, justify the use of this criterion in the current analysis.   (67–68)

The paragraph begins by directly relating to the counterclaim regarding the research method proposed in the paper. This claim has clear potential to weaken the paper's claim: If we accept it as is, doubt could be raised as to the validity and importance of the current study. The strategy pursued here is to explicitly put the counterclaim forward, and then to subsequently reject it. The rejection is carried out by a number of means, including the concession structure. This explicit strategy lends the text a dialogic character by bringing the rival claim (whether actual or potential) into the text.

It is notable that despite its explicit nature, Example (66) does not cite any specific author. However, although the claim is not attributed to a specific writer, it is clear from the wording that this is not a mere hypothetical claim, but rather one that has been asserted by researchers. This then is the voice of an unspecified member of the scientific community, as Thompson (1996) put it.

However, in Example (67) below, the rejected claim, which is worded impersonally ("One could ask"), is something that the researchers themselves thought of and tried to respond to (according to them successfully). From an argumentative standpoint, this can be seen as a way to submit (and reject) in advance claims that could come up in the reader's mind:

(67)   One could of course ask if the military service is what caused the maturation experience, or if the two to four years that elapsed are what made the difference. [...] We explicitly asked the participants in the study to attribute their retrospective assessments to the period of service and it is our assumption that that is what they did. Nevertheless, a self-report on the effect of the military service on areas of change that by virtue of their content are associated with army experience may be received with greater confidence. And indeed, we found a greater maturation effect in the instrumental direction in comparison to the expressive direction. The difference between the two directions in the intensity of the effect bolsters our confidence in the overall findings. A similar bolstering effect was also produced by the effects as differentiated by gender, rank and track.   (611)

In this case, there is a partial acceptance of the counterclaim with a rejection of the conclusions drawn from it. The partial acceptance of the counterclaim is expressed in the fact that it is presented in the form of an important and logical question, one that could significantly threaten the validity of the study's findings. This claim is immediately rejected by a description of the research method, which was aimed at overcoming the problem that was raised ("We explicitly asked the participants"). However, immediately afterwards, a further claim is presented ("Nevertheless..."), whose very mention weakens the response provided by the researchers. This claim serves, however, as a means to further strengthen the claim by means of the phrase "And indeed, we found...," which confirms it and establishes the connection between the findings and the conclusions. The two following sentences include bolstering phrases, and the entire paragraph presents a dialogue with a hypothetical skeptical reader, or reflects the researcher's inner dialogue with himself ("internal polyphony", Fløttum et al. 2006).

In the next example, the counterclaim is generally attributed to an identified but indefinite group of interlocutors ("the spokespeople of the advertising industry"). This claim is completely and explicitly rejected ("It is not reality"), showing that it does not represent the researcher's inner dialogue with himself. Immediately following, there is partial acceptance of another claim, one more limited and mitigated (reality is in fact not egalitarian) in the explicit structure of the concession that enables a more sophisticated and detailed rejection. The rhetorical force of the rejection is constructed in an interesting manner – first, by means of a short, decisive sentence followed by an elaboration of the same claim in the concession structure, which prevents a simplistic understanding of this statement as an empty slogan:

> (68)   It can *of course* be claimed, and this is what is generally done by the spokespeople of the advertising industry, that advertisements are merely a reflection of reality [...]. To this claim, one can respond: First, it is not reality. *Although* reality, in many senses, is far from being egalitarian, in the advertisements, the situation is much worse. [...] Second, it should be underscored that sexist messages reinforce the situation and consolidate prejudices.   (481)

Both examples above contain a modal phrase that weaken the counterclaim to some extent ("It can be claimed," "one can respond"), but on the other hand, the phrase 'of course' strengthens it to some extent. Regarding 'of course' (= *kamuvan*), it should be noted that it presents the counterclaim as self-evident, which makes it easier to accept it as having truth value; however, this also weakens its argumentative force: Since it is self-evident, it is obvious that the researchers have already considered the matter, and consequently, it is difficult to view it as posing a real threat to their claims.

Example (69) is taken from a paper about leaving ceremonies in the IDF. In the sentence preceding the concession structure, the authors quote claims that appeared in a study that downplays the social importance of ceremonies in today's society:

> (69)   The transition from one occupation to another or from one status to another during one's career is a turning point that generally involves a rite of passage. [...] But Giddens (1991, Chapter 3) maintained that the ceremony too, like the organizational framework, is rejected in postmodern society [...]. We believe that *even if* there is a growing tendency towards equality and dialogue, rather than towards hierarchy and ritual, rituals have not completely disappeared from the postmodern society, but have instead taken on a different form [...].   (104)

The concession structure is therefore enlisted in order to contend with an explicit claim made by a previous researcher. The schematic structure could be described as follows:

(70)   In the postmodernist world, ceremonies have lost their importance [and consequently, a study on ceremonies may not be important]; *however*, rituals have not completely disappeared, but instead have taken on a different form [and consequently, a study on ceremonies is important].

In the concession satellite, the authors express some agreement with this claim, but the wording moderates it so that the concession nucleus has no difficulty rejecting the conclusion that emanates from it.

Example (71) comes from a paper that deals with virtual organizations on the Internet. The satellite implies that the phenomenon is still too new to be investigated and that we are too close to it to allow for objective study. This is also a methodological claim that the author explicitly raises in order to reject.

(71)   We are at the epicenter of the information and media revolution (see Toffler 1992), and consequently we lack the necessary perspective to understand the social, organizational and personal significance of this new era. *Nevertheless*, and although we are unable to investigate this new phenomenon comprehensively and from the distance of time, we ought, already at this time, analyze the characteristics of these unique organizations, to assess their psychological and organizational significance and present the new issues that these organizations raise.   (513)

Let us now sum up our claims regarding the importance of the concession structure in scientific discourse.

First, the concession has argumentative value. After all, no writer seeking to persuade would directly weaken his argument by raising qualifications or counterarguments that work in the opposite argumentative direction from the one he wants to go in. How then will he contend with them? How can the writer prove that he is aware of the counterclaims or qualifications, and that he has already thought about them and decided to reject them? One way to do this is by embedding them in a concession structure. The counterclaim appears in the satellite, and alongside it, in the nucleus, appears yet another component that serves as a response to the counterclaim. The content in the nucleus can reject the counterclaim and prevent it from significantly weakening the argument.

Paradoxically, the final argumentative result of this process is to strengthen the argument. The reason, according to Roubrieux (1993), is that when using a concession, the speaker states in advance what may have been an unfavorable argument for his belief, and by so doing, he firstly eliminates a possible unfavorable intervention, and secondly reinforces the credibility of what is said in the nucleus. This occurs because the recipients are brought to believe that the speaker has already considered all possible objections, or at least all the important ones, and has rejected them all. By using concession, the author manages, on the one hand, to show that he is aware of the problem but, on the other, to reject the claim so as not to allow it to weaken his position.

The concession structure contributes to the dialogic aspect of the discourse from a number of directions. It is an obvious means to give voice to the "other" in the text. This "other" generally belongs to the discourse community, because otherwise, there would not be much point in making its voice heard and relating to its claims. Rarely, the "other" comes from outside the relevant discourse community, for example, claims attributed to the "media."

In certain cases, the claims in the satellite come from scientific publications that present theories and models, make hypotheses, offer interpretations, report on findings, opt for specific research methods, etc. These claims are attributed to specific members of the discourse community, and as such, their dialogicity is explicit, similar to the dialogicity typical of bibliographical notes. However, we have seen that the counterclaims can also be attributed to members of the discourse community whose names are not explicitly mentioned.

In other cases, the claims in the satellite have not actually been made, but are rather claims that the author assumes are likely to be raised against her own. In that case, the concession structure maintains a dialogue both with the reader, who may entertain such thoughts while reading, as well as with the entire research community. The dialogue with the disciplinary community in many cases has a social function too: Sometimes, knowledge shared by the members of the community is presented in the satellite, and its mention in the text serves to confirm the assumptions and consensus that they share.

Because the concession structure always contains an expression of agreement, it can help to create a sense of cooperation and membership within the community. The acceptance in the concession, as qualified as it may be, has a social value of cooperation. In doing so, the concession helps to strengthen the author's status in the discourse community. It may present her as a deliberate and cautious scientist, who has considered all the possible reservations that could be raised regarding her claims, and has addressed them; moreover, it may also contribute to the solidarity of the disciplinary community as a whole.

4.4 Inclusive *we*

4.4.1 First-person pronouns

The use of pronouns in a text clearly indicates its interactional nature. However, while the second-person pronoun is the most obvious means of directly addressing the reader, the first-person pronoun performs certain functions, some of which are directly related to the presence of the speaker in the text, with others being more complex.

The direct presence of the speaker in the text through the use of the first-person pronoun is a way for the author to establish his ethos. According to Aristotle, the *ethos* is one of the modes of persuasion of the speech. For Aristotle, the ethos is a reflection of the speaker's character, and he considers it to be the strongest means of persuasion (Aristotle 1982, 135a). Persuasion by means of one's character, says Aristotle, works when the words are uttered in such a way as to make the speaker appear credible. In the context of the scientific discourse, the moral character of the speaker is not generally a matter of interest, but what Aristotle called moral character can be thought of as the reliability and trustworthiness of the researcher, so that in this context, *ethos* can be identified with the researcher's credibility and standing in the discourse community. The characteristic that Becher (1981: 118) identified as the one shared by all researchers in all disciplines is the desire to earn a good professional reputation in the eyes of the members of the disciplinary community. We assume that the stronger the researcher's ethos, the greater the chances that his new arguments will be favorably received.

Numerous scholars of academic discourse have addressed the subject of the authority and the way to attain it by means of a written text. Scientific persuasion may be considered a result of a balance between two dimensions of authority: On the one hand, it is related to displaying personal qualities of reliability, and on the other hand to the ability to speak from the 'inside' as an in-group member (Bartholomae 1986; Cherry 1988; Hyland 2001b).

The author's membership in the community can be highlighted in a number of ways. For example, a survey of the professional literature can underscore his or her disciplinary authority: On the one hand, it helps to create a research space. This is a logical means to bolster a claim, and is consequently connected to *logos*. On the other hand, the very fact of demonstrating one's mastery of the research literature, and even more so – the ability to judge, critique and assess it, as well as to compare different approaches and positions, can serve as a means to strengthen the researcher's *ethos* by increasing his reliability as an up-to-date, discerning and critical professional.

Another way to establish the ethos is by employing the writing norms accepted in the disciplinary community. The use of disciplinary linguistic conventions, including impersonal, objective language, serves indirectly to establish the researcher's ethos. This is because it presents her as a professional who has mastered the conventions of the relevant discourse community. The use of the genre conventions per se is a valuable means of persuasion, and contributes to both the acceptance of the paper in the journal as well as of the new claims in the discourse community.

However, a paramount means for the establishment of the researcher's ethos is the use of the first person. The first person helps authors to establish their personal standing and to set their own work apart from that of others (Hyland 2001b: 217). The Anglo-American academic conventions influence scientific writing in other languages too and "encourage a conscious exploitation of authorial identity to manage the reader's awareness of the author's role and viewpoint" (Hyland 2002: 1111). The presence of the first-person pronoun in the structure "In this paper *we* report…" helps the reader to identify the main claim of the paper and its innovation (Myers 1992). Hyland (2002) found that writers choose to announce their presence where they make a knowledge claim. "At these points, they are best able to explicitly foreground their distinctive contribution and commitment to a position" (Hyland 2002: 1103). First person pronouns help writers create a sense of newsworthiness and novelty about their work, showing how they are plugging disciplinary knowledge gaps (Harwood 2005. For an overview of the research on pronouns in academic writing, see Harwood 2007: 29–30).

In a study based on interviews with scholars of political science (Harwood 2007), the interviewees were asked to comment on their and others' uses of the pronouns *I* and *we*. The analysis of their interpretations identified seven textual effects that these pronouns help to construct: (1) to make the readership feel included and involved in the writer's argument; (2) to make the text more accessible; (3) to convey a tentative tone and hedge writer's claims; (4) to explicate the writer's logic or method regarding their arguments or procedures; (5) to signal writer's intentions and arguments; (6) to indicate the contribution and newsworthiness of the research; (7) to allow the writer to inject a personal tenor into the text.

The researcher's presence in the text reflects a variety of roles and actions that he performs within his scientific work. Fløttum et al. (2006: 82) have identified four main rhetorical roles that the authors take on when referring to themselves by means of the first-person pronoun:

a.   Researcher – this role is typically taken when the author uses verbs which refer to the action or activity directly related to the research process, such as *analyze, assume, consider, examine, find, study* – which can be called 'research verbs.'

b. Writer – this role is taken when the pronoun is combined with verbs referring to processes involving verbal or graphical presentations, such as *describe, illustrate, present, summarize,* or processes related to text structuring and the guiding of the reader, such as *begin with, focus on, (re)turn to,* which can be called 'discourse verbs.'

c. Arguer – the author takes on this role when using verbs denoting processes related to position and stance, such as *argue, claim, dispute, reject,* which can be called 'position verbs.'

d. Evaluator – this role is taken when the author expresses emotional or evaluating reactions towards certain observations using verbs and constructions such as *feel, be content to, be skeptical about, be struck by.*

The terminology used by Fløttum et al. (2006) artificially assumes that the person whose name is signed to the paper is also the individual who actually carried out the research and wrote the text (Myers 1989: 4). As we know, this is not always so. In the research of certain sciences, it is not unusual for the names of researchers who did not actually participate in the study to be added, and it is reasonable in all disciplines that a paper signed by multiple individuals was not necessarily written by all of them. In any case, this issue is not the focus of our discussion. What is important here is that the author, the person who wrote the paper presents himself in the text in a number of different social roles: as a person who works in research, as one who is authoring the results of the research and as one making claims before the discourse community.

In a previous work (Livnat 2006), following Hyland (2000: 27, 2005: 184–185), I used a different classification for the acts in the text that might be assigned to the researcher: textual acts, physical acts and cognitive acts. This classification served me in the exploration of the functions of the passive voice in creating an impression of objectivity. There is a fundamental difference between presenting the physical activities in a way that is detached from the research and presenting cognitive activities in this way. In fact, it is not all that important who actually carried out the physical actions, and it is not at all unusual for these actions not to be carried out by the researchers themselves, or not by them alone. On the other hand, the strategy of presenting cognitive actions without attributing them to the researcher is an exceptionally artificial act, distinctive to scientific writing that has certain rhetorical functions in the genre, including the creation of the possibility of including the reader in the cognitive processes that underlie the study.

In the humanities, as well as in certain quantitative research in the social sciences, the act of investigation is often carried out during the writing itself, and in this sense, to "write a paper" means in fact to carry out a study. This fact raises the question of the differences between the roles of the *researcher* and the *arguer,* or between the *cognitive acts* and *textual acts*. However, in many papers, one may find utterances that are indicative that an author has made a distinction between these two actions. For example, an author may often write an apology of sorts that due to lack of space, not everything that was researched can be expressed in the paper, or an explanation that certain parts of the paper, (such as the explication of the terms that he uses) are presented only to help the reader better understand the research. In doing so, he himself defines the textual act as one that is distinct from the others.

4.4.2 First-person plural in Hebrew

The focus by Fløttum et al. on groups of verbs places the first-person pronouns (*I*, *we*), in the position of the syntactical subject in active sentences, at the center of attention. However, the morpho-syntax of Hebrew gives the corresponding pronouns a somewhat different status than in English, French or Norwegian.

First, in Hebrew, the first-person independent pronoun (both singular and plural) is syntactically obligatory only in the present tense. In the past or future tense, on the other hand, the morphological structure of a finite verb requires the designation of person, i.e. the verb necessarily includes a singular or plural agreement marker. See for example:

> Present tense: *ani bodek / anu bodkim* (I examine / we examine)

As opposed to:

> Past tense: *badakti / badaknu* (I examined / we examined)
> Future tense: *evdok / nivdok* (I will examine / we will examine)

The appearance of the agreement marker as a morpheme of the verb makes the use of an independent pronoun in the position of the subject superfluous, although it does not rule it out. And indeed, in written and formal Hebrew, in most first-person sentences in the past or future, the first-person pronoun does not appear as an independent component. Its occurrence in sentences in the past or future before a finite verb fills in written and formal Hebrew the pragmatic role of emphasizing and calling attention to the thematic role of the subject of the sentence. The syntactic structure is considered an extraposition and is "found mainly when the isolation of the pronoun is required by a context where the subject is changed or contrasted or added or described" (Goldenberg 1998: 177). This syntactic structure was found only four times in the entire corpus (including one in a footnote). Here are two examples:

(1) The care-giving questionnaire was developed in order to explore different patterns of providing help and care in the relationships between adults (Kunce & Shaver 1994). *We* (= *anu*) used (= *hishtamashnu*) it in order to examine the taking of responsibility and supervision over others in social relationships and the ability to show empathy among adolescents. (420)

(2) The debate over the ethics of teaching is generally of a theoretical and philosophical nature. *We* (= *anu*), on the other hand, decided (= *hechlatnu*) to explore how the teachers themselves perceived ethics and how it related to their day-to-day work. (442)

In Example (1), the authors use both the independent pronoun *anu* and the past-form verb (*hishtamash**nu***), that includes the plural agreement marker for first-person-plural (-*nu*). This kind of repetition emphasizes the contrast between the two sentences, i.e. the contrast between adults and adolescents as well as the different approach taken by the researchers in their use of the questionnaire, in comparison to its original intent.

In Example (2) too, the use of the redundant independent pronoun *anu* in addition to the verb *hechlatnu* calls attention to the thematic role of the researchers, as related to the contrast between two statements: "a theoretical and philosophical" debate as compared to theirs, which was a different kind of study.

The second complication is that first-person pronouns in Hebrew are affixed not only to verbs but rather to nouns or to prepositions. They may occur in syntactical roles other than as the subject. Here are a few examples:

> Possessive pronouns, affixed to nouns:
> Suffixed: *da'ati* / *da'atenu* (my opinion / our opinion)
> Independent: *hamodel **sheli*** / *hamodel **shelanu*** (my model / our model)

> Personal suffixed pronouns in prepositions and some adjectives:
> *oti* / *otanu* (me / us)
> *li* / *lanu* (to me / to us)
> *al-yadi* / *al-yadenu* (by me / by us)

Table 4 describes the distribution of these forms in the corpus.

Table 4. Distribution of pronoun in the Hebrew corpus

First-person pronouns		Agreement markers		Possessive pronouns		Prepositional complements	
singular	plural	singular	plural	singular	plural	singular	plural
2	68	23	231	5	73	–	32

Table 5. Singular and plural forms in single-author papers

Paper no.	No. of pages	First-person singular	First-person plural
1	24	5	6
7	18	10	6
10	24	1	2
15	7	6	–
16	15	3	7
20	17	1	3
23	15	–	6
26	17	3	1
27	20	1	3
28	19	1	–
29	15	–	23
Total	191	31	57

As can be seen from the table, the first-person pronouns in the subject position are less frequent in the corpus than as verbal agreement markers, i.e. when they are part of a verb-morphology. They are also somewhat less frequent than possessive and personals pronouns in prepositions and adjectives. All this poses a problem for an automatic search in a computerized corpus, because it is impossible to make a list of all the potential verbs and nouns to which the pronouns might be affixed.

Due to this complex situation, for the purpose of the current study I use the term *we* as a hyperonym that includes all forms of the first-person plural pronoun, in all possible syntactical functions, including agreement markers in the verb.

A central issue concerns the choice by a single author between the singular and first-person plural pronoun. The Hebrew corpus contains 11 single-authored papers. Table 5 shows the distribution of the choice between the singular and plural forms in these papers.

This small sample shows that in single-authored papers in the social sciences in Hebrew, the authors used the plural form almost twice as often as the singular form. In two papers, only the plural forms were used, and in two other papers, only the singular forms were used. More than two-thirds of the singular forms were verbs (23 out of 31 verb forms in all), and all were embedded in meta-textual utterances relating to the structure of the paper and the manner in which the claim was presented, in other words actual acts that put the role of the writer into practice. For example:

(3)   This is the question that *I will expand* upon.   (10)

(4) At the beginning of the paper, *I touched* on questions related to the boundaries of childhood and the division of age groups.   (376)

(5) As *I have shown*, Israeli law determines many levels of rights and responsibilities [...].   (378)

(6) *I will note* in the description only those characteristics that contributed more than 2% to explain the diversity [...].   (626)

The most significant finding is that not even one of the actions described in the first-person singular denotes a physical action. The explanation for this is that the Method section is the most impersonal part of the paper. Here the author's addressing of the conventions of impersonal writing are most clearly evident, and in which the distance between the researcher and the study receives maximum expression (Livnat 2006).

First-person singular can also be found in a number of hedges that in Hebrew include the pronoun:

(7) *To the best of my knowledge* (= *lemeitav yedi'ati*), no study has been conducted to explore the influence of the intent [...].   (664)

(8) Whereas Costa and Garmston developed their model for behavior among peer teachers, *it appears to me* (= *nir'e li*) that it can be applied by framing dialogue situations between teachers and students [...].   (633)

The pronoun *I* in the position of the syntactical subject appeared only once in the body of the text (and once in the footnotes):

(9) *I am* hopeful (= *ani tikva*) that the findings of this study will arm those setting out on this struggle with empirical data and represent the beginning of a regular follow-up study.   (482)

The phrase *ani tikva* (lit. *I am a hope*) is a high-register idiomatic collocation, which appears in the final sentence of the paper. The sentence is worded figuratively ("arm"; Hebrew: *lechamesh*) and serves as a kind of coda (Labov 1972) wielding a certain force of expression. This is also the only paper in which a clear personal opinion and position are expressed. This is what Hyland calls an *attitude marker*; its use here is related to a strategy of foregrounding the author's voice at those points in the paper that the author seeks to emphasize. First-person pronouns are a natural form of expression through which the voice of the researcher is reflected. By using them, he can demonstrate responsibility for his claims, underscore them and send a message of reliability and authority that will be beneficial in establishing his standing in the disciplinary discourse community. The first-person pronouns have then a number of rhetorical functions: They help to highlight the self, increase the responsibility of the writer to his claims, thereby contributing to his ethos. They can benefit the researcher's ethos also by underscoring the fact that he is a valuable member of the disciplinary discourse community.

Another matter related to the author's presence in the text by means of the first person is that of gender. In Hebrew, verbs in the past and future tense, in the first and second person, are not marked for gender; however, in the present tense they agree with the gender of the subject, both in the singular and plural. In addition, the masculine is the unmarked form, in the sense that it also refers to a group of people that can be made up of both men and women. Thus, the use of the masculine may engender a generic interpretation, signifying people whose gender is unknown to the speaker. Consequently, in order for the plural form to be ambiguous so that it can include the readers too, it should be in the masculine [grammatical] form. On the other hand, the plural in the feminine form is unambiguous, and is therefore used only when all the researchers are women. My corpus contains seven papers that were written by two women, and I will present some examples of feminine first-person plural forms excerpted from them:

(10)   In order to better understand the personality and emotional structure of the parent-child, *we referred* (fem.) to attachment theory [...].   (416)

(11)   According to this perspective, *we do not consider* (fem.) the Zionist ethos to be something that is self-evident and *we are not interested* (fem.) in exploring whether or not the immigrants indeed fulfill it.   (573)

The markedness of the feminine forms in Examples (10)–(11) lends a strong presence to the women researchers in the text, in contrast to the corresponding masculine forms, which as noted are unmarked and consequently more open to interpretation.

When the researchers are exclusively women, the use of a masculine form is also unambiguous: In this case, it is clear that it is inclusive and that its referent is not the researchers alone, but also the readers too, as in Example (12):

(12)   In these schools, *we would expect* (masc.) to see flights of imagination and inspiration in the want ads.   (258)

The Hebrew phrase *hayinu metzapim* is the masculine form, as opposed to the feminine *hayinu metzapot*. By the use of the masculine form, the cognitive act of "expectation" (= prediction) is presented as if it is not attributed to the women researchers-authors of the study only; rather, it is shared with the reader too. When the expectation is not realized (which is implied by the phrase "we would expect"), the reader is supposed to be just as surprised as the researchers were. The use of the masculine first-person plural form enables the reader to be involved in the cognitive processes underlying the study: raising hypotheses and examining them in relation to the findings.

I found this masculine form in two other papers written by women researchers:

(13) In this case too, *we are witness* (masc.) to a phenomenon that has been gaining momentum in recent years.   (509)

(14) Since the 1970s, *we have seen* (masc.) increasing criticism in the area of secular culture too […].   (585)

In these cases, the plural form (*anu edim*) is interpreted as not only referring to the researchers and the readers of the paper, but also as relating to the entire scientific community or perhaps even members of Israeli culture in general, who serve as witnesses observing the described phenomena. The possibility to expand the reference of *we* to different groups is of great importance in the scientific discourse, and I will discuss it in the next section. An important point to remember, however, is that the referent of *we* might not be fully determined by the author. In Examples (13)–(14) it is interpreted, among other clues, based on the incompatibility between the masculine form and the gender of the researchers (which the readers can figure out because of their first names). In other cases, this interpretation might depend on the reader's interpretation of the text (Fløttum et al. 2006:97).

4.4.3 Inclusive *we* as dialogue

Whereas the singular form unambiguously marks the researcher, the plural form can refer to a variety of referents: a single researcher or a group of researchers who collaborated on the paper; it could include the reader, refer to an entire community in a social or cultural context, or even refer to human beings in general (Martín-Martín 2005:143; Fløttum et al. 2006:109). Many scholars have discussed the pragmatic complexity of first-person plural pronouns. Fløttum et al. (2006:95) propose to break down the complex picture into two main pragmatic issues: the question of reference and questions of role relations and rhetorical motivations.

From the referential perspective, the main division is between exclusive and inclusive uses. Although many authors have attempted to make finer and more intricate classifications (cf. e.g. Loffler-Laurian 1980; Wales 1980, 1996; Rounds 1987; Mühlhäusler & Harré 1990; Kuo 1999; Fortanet 2004; Fløttum et al. 2006), the exclusive/inclusive distinction remains the clearest and most solid one.

The exclusive use of *we* excludes the addressees and thus may refer to multiple authors of an article, or to refer "metonymically" (Fløttum et al. 2006: 95) to a single author. In the following examples taken from single-authored papers in Hebrew, it is clear based on the context that the referent of the pronoun is the particular single author whose name is signed to the study:

(15)   In Table 1, *we compared* the proportion of attendance by the sexes based on
       different indices.   (472)

(16)   The data in *our study* bolster their claim [...].   (16)

(17)   The result that was obtained in *our analysis* is almost completely compatible
       with [...].   (679)

The explanations for the use of this convention are numerous and far from unequivocal. The plural form can express the author's modesty and self-effacement in that it minimizes the presence of the writer in the text. It can also send a message of objectivity and strengthen the argument, since "scientific knowledge is supposed to be taken as universal; therefore any implication that a belief is personal weakens it" (Myers 1989: 14). However, as Hyland (2001b) suggests, "use of the plural is only partly explained by patterns of authorship [...]. It is not always the self-effacing device it is sometimes thought to be. Pennycook (1994: 176) for example, observes that "there is an instant claiming of authority and communality in the use of *we*." Examples from single-authored papers "suggest how writers can simultaneously reduce their personal intrusion and yet emphasize the importance that should be given to their unique procedural choices or views" (Hyland 2001b: 217).

Inclusive use of *we* includes one or more of the addressees, whether they are hearers in a spoken interaction or readers of academic articles. The rhetorical motivations for the various uses of inclusive *we* lie at the center of our concern here.

The replacement of *I* by *we* is one of the means through which "a speaker endeavors to get his audience to participate actively in his exposition, by taking it into his confidence, inviting its help, or identifying himself with it" (P&OT 1958: 178). By using this linguistic resource, "points of view are attributed not only to the authors and other researchers mentioned in the bibliographical references, but to the reader and to the disciplinary collective of researchers" (Fløttum et al. 2006: 110). It is then an extremely important dialogic element. Readers are most commonly brought into the text as discourse participants by means of the use of the inclusive *we* (Hyland 2001: 557).

The action of attributing the referent to the inclusive *we* is a complex interpretive action, as well as a description of its rhetorical function in every context. Hence, I do not plan to describe the full range of possibilities that arise from the entire corpus of authentic texts, categorize them and explain the ambiguous cases, a task that will emerge as exceedingly complex and virtually impossible (Fløttum et al. 2006: 101). My intent is far more modest: to present two functions of the inclusive *we* that appear to me to be the most important ones and to distinguish between them: addressing the reader and addressing the discourse community. These two functions are based on Myers proposal (1989: 4) to replace the simple model of the relationship:

Speaker $\longrightarrow$ Hearer

with a messier relationship, one that is more relevant when analyzing academic discourse:

Author as writer $\longrightarrow$ Readers
Author as researcher $\longrightarrow$ Researchers

It is true that the reader of academic articles is usually a member of the discourse community, and one might say that by addressing the audience of readers, the author is in fact addressing the discourse community. Nevertheless, I consider it important and worthwhile to distinguish between an appeal to the reader and inviting him to share the reading situation, on the one hand, and the mention of the discourse community in general terms, on the other. Even if it is not always a simple matter to ascribe a particular example to one category or another, I view the fundamental distinction between them as important. Moreover, the community that the author is addressing is not always the disciplinary scientific community, and may also be a larger group, as we will see.

(a) *Inclusive we for reader's participation.* The appeal to the reader in reading situations and inviting him to be part of it is related to the design of the text when leading to what Bakhtin called *responsive understanding* (see Section 2.1 above). Example (18) below is presented as part of a discussion following excerpts of interviews that cited five types of responses to a particular question:

(18)   The five voices *that we have heard* expressed agreement that Israel is the
        national home of the Jews [...].   (583)

Because the interview excerpts were categorized and discussed in the paper so as to reflect five different positions, the assumption of the researchers is that not only they "heard" the five voices, but that the readers did too. The form "that we have heard" invites the reader to form an impression similar to the one that the researchers formed. The same is true for the phrase "we hear" in Example (19), which follows a series of quotes from an interview with the person under discussion:

(19)   In his words, *we hear* contradictory messages that are indicative of his ambivalent state.   (119)

In general terms, the activities that the reader is invited to share with the author by means of the first-person plural pronoun are activities on two levels. On the overt level, they are activities related to following the structure of the paper, organizing the subjects in it and connecting its various parts, for example:

(20)   So far, *we have seen* that the ethnic perception of identity is strong […]. (16)

This utterance serves to sum up one idea and enable the transition to a new one. The summary is presented as one that is not one sided or biased towards the author. This is the case in Example (21) too, which is taken from a paper that analyzes newspaper texts. First, the paper presents a text excerpt for analysis, followed by Example (21), which serves as preparation for the linguistic analysis that immediately follows:

(21)   *We will examine* a number of the elements of the stylistic phrases, their components and the impression they make […].   (159)

Sometimes formulations such as these appear when an unnatural structural process is carried out, such as a return to something that was discussed earlier:

(22)   If *we return* to a number of the figures noted earlier, we will see that Lieutenant Colonel Yehuda […].   (121)

The role of meta-textual utterances of this kind is to help the reader find his way around the structure of the paper; however, the use of the first-person plural pronoun also enables the reader to more tangibly participate in the process of putting forth the arguments. It is "a strategy that stresses the involvement of the writer and the reader in a shared journey of exploration, although it is always clear who is leading the expedition" (Hyland 2001a: 560).

This strategy is even more significant when activities on the deeper level are involved: the cognitive activities that are related to the raising of hypotheses, creating generalizations, drawing conclusions, etc. Through the choice of the plural form, the author shares the cognitive processes that advance his claim with the reader, so much so that the reader feels as if he carried them out himself. Let's look at the following example:

(23)   In light of this, *we would expect* the firstborn children – more than the other children – and girls – more than boys – to have a greater tendency to adopt the role of the parent-child.   (416)

The cognitive act of "expecting" is one that the researchers carry out, but the inclusive plural form enables the reader to participate in it too as he follows the logical progression of the argument. In Example (24), the author asks the reader to accept the categorization upon which he bases his arguments, in the context of the discussion of the findings. To this end, he shares the cognitive process of accepting the definition with the reader, along with the continued action in accordance with this definition:

(24) According to Cherniss, the feeling of self-efficacy on the part of the manager could be made up of the task area [...], the area of reciprocal relations and of the organization [...]. If *we accept* this expanded definition of the concept of self-efficacy, *we can* classify the pressures of the job, as consolidated in this study, based on three categories [...].   (239)

This sharing of cognitive processes with the reader is especially salient in those studies in which the *responsive understanding* is not guaranteed. I will demonstrate this from a paper that was published in a social science journal, in which the author offers a model he developed on his own to identify groups. The model is a mathematical one and it is evident that the author is making considerable effort to explain it to the members of the discipline, who may not be accustomed to reading material of this kind. To accomplish this, he uses a large number of verb forms in the first-person plural to lead the reader one step at a time along the complex cognitive process that needs to be carried out in order to appreciate the advantage that his model has over others and the truth of his claim. Here are some of them:

(25) *Let us go back and look* at the group of monks in the monastery. Group A-{16}, the "group of Young Turks for Monk 16" is a lattice $k = 4$ network (definition 3), because each individual in it has a reciprocal connection with at least two other members of the group. If *we look* for lattice $k = 4$ networks, *we will find* network A, but *we will not find* group B, the "opponents," because individual 5 is reciprocally connected only to one individual from the group [...]. On the other hand, *we will find* another lattice $k = 4$ network [...].   (699)

(26) A network is a group of individuals who prefer to connect with one another than with individuals outside the network [...]. The intensity of this preference can be expressed numerically with the help of a segregation matrix index (SMI). From this it follows that when *we review* all the various possibilities of groups in the social network and *we discover* the groups for which the SMI obtains the maximum local value, *we can define* these groups as networks.
                                                                                      (691)

Utterances of this kind construct dialogicity with the reader; their purpose is to help the reader follow the logical structure of the author's argument and the cognitive processes that support it. In addition to the help they provide the reader, the feeling of partnership that the use of the plural creates also has a rhetorical effect. The reference to the reader "sends a clear signal of membership, textually constructing both the writer and the reader as participants with similar understanding and goals" (Hyland 2001:558). The reader is assumed to be affected by this participation. Fløttum et al. believe that the reader is engaged in the research process "and is assumed to agree: What he or she has done, obtained and observed together with the author will more likely be accepted as true, relevant and reasonable" (2006:97).

We must not forget, however, that this is essentially a persuasive strategy. While trying to predict and respond to their readers' line of thought, the authors "are also trying to encourage particular reactions to their argument – specifically, to secure their agreement" (Hyland 2001a:558). In my view, the feeling of sharing and involvement in the cognitive process can also cause intellectual enjoyment on the part of the reader at reading the paper and at the manner in which the new argument is structured. This enjoyment serves to increase the reader's positive stance toward the paper and the arguments presented in it, thereby increasing the chances of their acceptance.

(b) *Communal we.* The vague reference to the disciplinary community is an important function of the inclusive *we* (Fløttum et al. 2006:97; Myers 1989; Kuo 1999). Through the use of the first-person plural, the author of the academic paper can not only involve the reader in the reading process, as we saw in the previous section, but also reference his direct discourse community, which is grappling with questions similar to the ones that he himself is contending with in his paper. Let's look at this example:

(27) Nevertheless, it is notable that there is a problem in estimating the potential for the flare up [...]. It appears that *we do not have* the tools to enable us to precisely predict if collective activity will take place at all and to what extent.

(24)

Rhetorically speaking, a first-person form of this kind is not related to highlighting the speaker in the text, but rather his membership in the scientific discourse community and the fact that he shares its scientific endeavor, its goals and the tools to attain those goals. The claim that "we do not have the tools" (Hebrew: *eyn beyadeynu kelim*) points to the problem that the author and other researchers share, all those that take an interest in the subject under discussion. In the following examples too, the use of *we* points to the difficulties that researchers working in similar fields encounter:

(28)   The concept of the network in the social network appears simple; however, when *we* try to construct a process to identify networks, *we* encounter mathematical difficulties.   (700)

(29)   *We* are facing a new kind of organization, and the accepted concept of "organization" is increasingly losing its meaning.   (529)

(30)   *We* are at the epicenter of the information and media revolution (see Toffler 1992), and consequently, *we* lack the necessary perspective to understand the social, organizational and personal significance of this new era.   (513)

The writer in these cases presents not only his perspective as a researcher, but also one that is not attributed to a specific "other," and therefore enables other researchers to identify with it. The author uses the communal *we* here in order to represent the voice of the community or of an "unspecifiable other" (Thompson 1996:510). The forging of solidarity with the discourse community by means of the communal *we* can be achieved by presenting the goal of the writer as one shared by the disciplinary community, which, like the writer, seeks to increase its knowledge. This is evident in the following example:

(31)   Comparing the answer to this questionnaire with the answer to the question regarding the participants' land of birth [...] can *help us to learn* about the differences between "objective" futures [...] and subjective ethnic identity [...].   (11)

It is not always clear if the group that the communal *we* refers to is limited to the members of the disciplinary community or is perhaps broader. In the following example, it could be claimed that those that "may be witness" to the phenomenon under discussion are not only researchers coming from the same discipline; in fact, any person living in the relevant society during the relevant period could be witness to these phenomena. At the same time, the researchers are the only ones to observe the phenomena through the prism of concepts such as "formal group," "informal group" and "weak ties."

(32)   *We may be* witness to the rise and strengthening of numerous formal and informal groups of "weak ties" (Granovetter 1983), including ties created with the help of the electronic media.   (374)

A similar question arises in the following example:

(33) *We do not know* how these changes will affect parent-child relationships and children's development. (376)

Who does "we" refer to here? While it is true that no one knows what influence the changes that are mentioned will have – it is not only the researchers in similar areas of study that do not know, but no one at all knows – at the same time, the researchers are the only ones who are taking an interest in this question. The author here is then addressing her discourse community by raising issues and questions that interest it. For the sake of comparison, let's look at other examples taken from the same paper:

(34) There is no doubt that significant changes have occurred *in our time* (Hebrew: *bizmanenu*) in the growth and maturation process of young people. (379)

(35) These are trends that not only *our children*, but also *we* as adults need to adjust to. (366)

In these cases, it is clear that the "community" that is mentioned by means of the plural forms is not necessarily the research community, but rather the members of a particular society and culture at a particular point in time.

As noted, the papers in the Hebrew corpus are all taken from the social sciences, a field that by nature discusses social and cultural contexts. For example, in a paper that deals with the roles that adolescents adopt in the family, the following sentence appears:

(36) Thus, the findings of the interviews […] show that the children […] try to advise, comfort, calm and, especially, please their parents, far beyond the level that we are accustomed to in *our society.* (433)

The pronoun in "our society" (= *tarbutenu*) clearly includes not only the researchers and readers, and not only the members of the scientific community, but all the members of that particular culture, apparently Israeli society or perhaps Western civilization as a whole. Other papers in the corpus, which describe broad-based social-cultural phenomena that are not necessarily unique to Israel, tend to use this type of plural form too. In a paper dealing with virtual business organizations on the Internet, 8 of 25 uses of *we* are in a cultural context:

(37) […] The innovations of modern technology place *us all* in a wonderland in which *we* must rapidly adjust to the new dimensions of time and space, which are fundamentally different from the reality to which *we are accustomed.* (512)

Example (38) is an interesting case; it comes from a paper dealing with creativity and ways to measure it:

(38)   Another example is the well-known puzzle: "How can you use six matches
to build four equilateral triangles?" The solution to the puzzle is to build a
pyramid [...]. The rule belongs to a new type, since *our knowledge* of the
"matchstick puzzles" creates a context that involves work on two dimensions,
and that is the expanse where we are most likely to look for the solution.
(41)

The referent of the pronoun in "our knowledge" (= *hekerutenu*) is approximately
"the group of human beings familiar with puzzles involving matches."

Martín-Martín (2005: 142–143) proposes to divide the category of inclusive
we forms into two subcategories: (a) cases where the pronoun refers to people in
general; (b) cases where the pronoun refers to the members of the disciplinary
community. It is not surprising that he found that the second subcategory is much
more frequent in English and Spanish abstracts. This is the kind of Inclusive *we*
that I term the communal *we*, and the significant rhetorical role of pronouns of
this kind is closely related to dialogicity, as Martín-Martín himself explains:

> This function may represent an attempt on the writer's part to signal their de-
> sired membership in the discourse community by displaying knowledge of the
> facts and opinions that are generally accepted by the disciplinary group. This use
> reduces the distance between writers and audience and emphasizes solidarity,
> indicating shared knowledge between the writer and the reader, and a presup-
> position of the writer's acceptance by the discourse community.   (2005: 143)

I propose then to view the communal *we* as a phenomenon that has special value
in the academic discourse, and to distinguish between it and the inclusive *we* that
addresses the reader.

To sum up the discussion of pronouns: The masculine first-person plural
forms are the generic form in Hebrew and consequently, can express different
voices. They can express the voice of the researcher himself, both the individual
researcher who chooses to use the plural form for various reasons, as well as a
group of researchers who are responsible for a paper. However, like in other languages, the first-person plural forms can be inclusive and include the reader too.
When using it in this way, the author addresses the reader as part of a dialogue,
involves her in the cognitive processes that underlie the text and guides her along
the paths of the argument. All of these are rhetorical means that act to help the
new claims to be accepted.

The inclusive *we* can also express the voice of the community. It may be the
discourse community, that is the narrow research community to whom the text is
directed, but it can also be a larger one. Rhetorically speaking, the dialogue with
the discourse community, which is carried out by means of the communal *we*, is
of special importance. This dialogue enables the researcher to express solidarity

with the community, establish his social status in it, along with the status of his text as one having value for the shared endeavor of the community.

4.5 Questions

4.5.1 Direct and indirect questions

Questions are a major dialogic component. Bakhtin (1986:72) notes that asking questions is typical of persuasive genres. They are "the strategy of dialogic involvement *par excellence* [...] working to create rapport and intimacy" (Hyland 2002b:530). It is "one of the most common ways that writers project the perceptions, interests and needs of a potential audience into their unfolding argument" (531).

The discussion of the status of questions as a dialogue component means that we must first distinguish between direct and indirect questions. A comparison of the genres shows that direct questions are typical of spoken discourse and interpersonal interaction. Biber et al. (1999:211) found in the 40-million-word Longman corpus that direct questions were found to be fifty times more common in conversation than in academic writing.

Indirect questions, on the other hand, are considered a distinct feature of written discourse (Chafe 1985). Consequently, we do not expect to see a significant presence of direct questions in academic papers, in comparison to indirect questions.

As anticipated, indirect questions are more common in the corpus of Hebrew papers too, especially the indirect question positioned as a relative clause of the noun "question," as in this example:

(1) The question is asked whether the linear context [...] is the result not of the multiplicity of identities, but rather of the differences between the various combinations of these identities.   (212)

The use of the noun 'question' is a form of compensation for the indirect wording, i.e. for the absence of a genuine question form, and it highlights the status of the clause as a "question." Less frequently, the clause is positioned as an object clause of the verb 'ask' in different inflections:

(2) [...] When analyzing the results, *we ask ourselves* if the significant changes in this matter can be defined as a natural phenomenon. [...] On the interpersonal level, *we ask ourselves* how the death of a loved one affects relationships [...].   (282)

In this case, the speaker is present in the text due to the use of the first-person pronoun; however, because the question is a reflexive one – of the author to himself – the interaction with the addressee may be defined as weak.

A loose syntactic construction that mixes the indirect and direct wording can be found in my corpus in a number of places:

(3) These studies raise the question: Under what conditions can shared learning encourage the resolution of mathematical problems presented in different problem situations?   (664)

(4) One of the central questions of occupational insecurity is: How does occupational insecurity affect workers and their outlooks?   (488)

The loose syntactical connection (by means of a colon) enables the writers to use the question mark, which needs to be read with a certain intonation, and thereby underscore the "asking" nature of the utterance. This is the case in Example (5), in which there is a mixture of the two structures:

(5) For this reason it was decided to present not only a description of the teaching strategy in the current study [...], but also to try to find out: what affects the preferences of teachers regarding the ways of coping?   (620)

However, contrary to anticipation, and like the findings from the two main studies that examined the distribution and function of questions in academic discourse in English (Webber 1994; Hyland 2002b), quite a few direct questions were found in the Hebrew corpus too (77 direct questions – an average of more than two per paper). This fact bolsters the definition of the academic paper as a persuasive genre and points to its dialogic nature. The examples I have presented thus far help to understand the unique nature of direct questions that end with a question mark. The direct formulation increases the dialogicity of the text and creates a feeling that the addressee is being directly addressed, even if he is not present.

In his discussion of direct question in academic papers, Hyland (2002b) distinguishes between "rhetorical questions" and "real questions." He does not provide a precise definition of these two categories, but he appears to include in the category of rhetorical anything that is not "real questions," the kind of questions which Searle (1969:66) defined as an attempt on the part of the speaker to elicit information from the hearer. Thus, this category might include questions that others relate to as *expository questions* (used to introduce a new topic), *deliberative questions* (used to make decision by choosing the best alternative), *speculative questions* (used to put forward hypotheses) (Illi 1994:39) and *conductive questions* (questions that convey a questioner's expectation and preference for a given answer) (Piazza 2002:510).

This definition is very broad, and consequently it should come as no surprise that real questions are fairly rare in his corpus of English papers. On the other hand, over 80% of the questions were rhetorical, "presenting an opinion as an interrogative, so the reader appears to be the judge, but actually expecting no response" (Hyland 2005: 185–186).

Hyland also included in the category of "rhetorical questions" the strategy according to which the author asks a question and then answers it himself. According to his findings, this textual pattern is prevalent in the "soft" sciences. Here is an example from the Hebrew corpus:

(6)   If so, what is a child according to Zuta? It is in fact a young person who is invited to take a quick glance at selective and distorted information behind the scenes [...].   (537)

In this case, the answer was supplied explicitly and briefly immediately after the question was asked, and in this, the author is "simultaneously initiating and closing the dialogue" (2002b: 549). In this way, the question, which has the ability to provoke thought and involvement in the text in the reader, in fact does not make this possible. However, in other cases, the answer provided by the author is long and complex, and consequently does not immediately close the dialogue. Let's look at an example from the Hebrew corpus:

(7)   The weaknesses of the system are the target of criticism shared by most of the public. But why did the students suddenly offer to say what everyone else was thinking, to voice this genuine outcry, one that does not offer any new facts? One possibility is that they were working under the illusion that they would be able to sway the masses to support their justified cause [...]. Another less likely possibility is that the opposition exploited their distress [...]. A third possibility, one that is even less likely, is that it was the government and treasury that viewed the strike as beneficial [...]. My opinion is that they went on strike due to the needs and inner dynamics of campus life.   (318–319)

In fact, the question here serves as an introduction and justification for a long series of alternative explanations, at the culmination of which the author presents his own opinion. In response to the question, the reader might entertain explanations of his own, some of which might resemble those raised by the author. This creates a hidden dialogue between them. By means of the strategy of raising various explanations and rejecting them, the author is leading the reader step by step to the answer he considers most suitable. At the same time, from a practical standpoint, the author provides the answers to the questions in the text. Many accepted definitions of rhetorical questions do not include this type of question.

Two crucial features of rhetorical questions are mentioned by Illi (1994) among the five distinctive features she proposes: (1) There is a discrepancy between the interrogative form of the rhetorical question and its communicative function as a statement; (2) There is a polarity shift between the rhetorical question and its implied statement, i.e. a question in the affirmative usually implies a negative answer and a question in the negative usually implies an affirmative answer (Illi 1994: 45). This is an example from my corpus that reflects these features:

(8) In this way, paradoxical situations can be created – [...]. Parents can force their children to become religious, but are not permitted to convert their children to another religion after the age of 10 without their permission. These children do not yet have criminal liability. Does converting to another religion require a lower level of maturity than vandalizing property or causing harm to another person?   (370)

The adjective "paradoxical" presents the author's negative view in advance regarding the logic of what will follow. Consequently, the answer to the question is self-evidently interpreted as a negative one. Questions such as this are very rare in my corpus. Example (9), on the other hand, is more ambiguous because the position of the authors is presented (in the final sentence) in a more moderate and qualified fashion:

(9) Courage [...] is revealed to be a rare quality. However, can teachers form a moral leadership having strength and influence without sources of authority to give them legitimacy and support? Can we expect teachers in their current public status to challenge accepted states of mind? In face of the demand of teachers to demonstrate courage and rise above considerations of convenience, their considerable dependence on the social and organizational reality in which they work cannot be ignored.   (463)

The two-term formulation of the last sentence, which creates a balance between the two positions, makes it possible to view the questions not necessarily as rhetorical, but rather as real questions that allow the reader the possibility of providing his own answers to them.

Due to the problematic nature of the term "rhetorical questions" and the multiple definitions of it (see Illi, Chapter 3), the distinction that I will use in the following section will be between real questions and non-genuine questions. Real questions are those to which no answer is provided in the text; non-genuine questions are those that are answered in the text, either explicitly or implicitly. I will argue that these two types of questions are distinct in their dialogic nature: Non-genuine questions address the reader in the specific context of the paper, and real questions appeal to the discourse community in the broader context of the shared scientific project.

4.5.2 The rhetoric of questions

Following Adams Smith (1987), Swales (1990) and Webber (1994), among others, Hyland (2002b) points to seven central functions of questions in research articles, which I have arranged here from the most to the least frequent: (1) organizing the discourse; (2) conveying a claim; (3) framing purposes; (4) expressing an attitude or evaluation; (5) suggesting further research; (6) establishing a research niche; (7) arousing interest.

Hyland provided an exhaustive rhetorical analysis that offers examples of each of these different functions. Consequently, I will present here some of these possibilities in brief, and focus on the different dialogic aspects of the questions, with an emphasis on the basic distinction that I have proposed between appealing to the reader and appealing to the discourse community.

(a) *Non-genuine questions directed to the reader.* Questions clearly directed to the reader are those that appear in the titles of papers. Presenting a question as a title or in the title contains a promise on the part of the author that the paper will supply the answer, and consequently stimulates interest, or even an emotional response on the part of the reader. In Hyland's view (2002b: 538), the role of the title formulated as a question is to contend for the potential reader's attention. In an environment in which the reader has numerous options, one has to make sure that he will read this particular paper. This also explains why Hyland's corpus does not contain titles of this kind in the "hard" sciences, but rather only in the "soft" ones. In the "hard" sciences, in which the structure of knowledge is more compacted, the paper does not generally need to draw the attention of readers. In the "soft" sciences, in which the structure of knowledge is less fixed, the possibilities open to the reader are indeed far greater and are not binding to the same extent. Hyland does not explain why a title formulated as a question will attract the attention of the potential reader more than one that is worded differently, but it is reasonable to assume that one of the reasons is the infrequency of such titles. My corpus contains two papers that have questions in their title:

(10)   Virtual communities: A new social structure?   (298)

(11)   To stay in the country or leave? – The challenge of the Zionist ethos in immigration stories   (571)

At first glance, the wording of Example (10) appears to be that of a research question, when in fact, this is not the case. The thematic noun phrase "virtual community" indeed expresses the general subject of the paper. However, in their paper, the researchers do not answer this question, but instead assume that this is indeed a new social structure and consequently is worthy of research. This shows that the authors do not consider the question in the title to be a question at all. In Example (11), the question in the title is not asked by the researchers, but is rather one that the interviewees whose stories are told struggle with.

Why then are these questions asked in these cases? It would appear that the rhetorical explanation that Hyland offers provides an answer: It is a different, stimulating and attractive formulation that draws attention. This appears to be especially true regarding to the question "To stay in the country or leave?" whose formulation is somewhat simplistic and unscientific, and consequently draws attention on the background of what readers have come to expect from conventional scientific language. In the text itself, this same question is formulated, as might be expected, in the form of an indirect question:

(12)   The question of whether to remain or leave the country is then one of the ultimate tests that the Zionist ethos demands of the Jewish residents of the State of Israel.   (572)

It is interesting that the English table of contents of the volume doesn't include a question either. Instead, with no attempt to vie for the reader's attention, the title is worded thus:

(13)   Immigrants challenge the national ethos: Jewish-Russian students deconstruct Zionism.

A considerable proportion of the questions in the academic paper are related to the dialogue with the reader. The author can use them to draw the reader's attention and direct him to specific points in the argument. The assumption is that the question has the power to draw the attention of the addressee because is seemingly directed at him and wants him to think about it. In the academic discourse one can find questions "where writers seek to explicitly establish the presence of their readers in the text: inviting engagement and bringing the interlocutors into a discourse arena where they can be led to the writer's viewpoint" (Hyland 2002b: 530).

Questions play an important role in organizing the discourse by cutting short the sequence of the text and helping in the transition from one subject to another (Montgomery 1996: 38). Questions such as these support the progression of the discourse and the argument. Let's look at this example from the corpus:

(14) We have seen so far that most Mizrahim have an ethnic identity [...]. We have
      also seen that there is a growing sense of collective discrimination [...]. But
      what about Jameson's third condition – a sense of agency that helps in resolv-
      ing the problem? In order to answer that question, we need to examine the
      connection [...].   (18)

Some of the questions that appear in the academic discourse represent the category of "research questions", a conventional term that relates both to the researcher's
cognitive act as well as to a structural part of the paper. This term implies that at
the heart of the scientific work is a question that the researcher is supposed to try
to answer in the paper. The entire text can appear to be framed in a question-an-
swer or problem-solution structure (Swales 1990: 137–140), i.e. the importance of
the study lies in the answer that it provides to a particular question or the solution
that it offers to a particular problem, which also may be formulated in terms of
a "question." Hyland (2002b: 540–541) believes that one of the functions of questions in academic papers is to frame the discourse: In the initial paragraphs of
the paper, the authors present the question that they plan to answer, and as the
paper progresses, they expand upon it and formulate their answer in regard to it.
From this it follows that the formulation of the "question" or "questions" in the
academic papers is not only expected, it is required.

In my corpus of papers from the social sciences, the category of "research
questions" as a structured section was not found in every paper. Even in those papers that did have this section, the research questions were not always formulated
as a question. Only in three of the 30 papers did I find direct questions formulated
as such under the heading "research questions." However, in other papers, direct
questions representing the research questions could be found in different parts of
the text, not necessarily under the heading of "research questions." These questions were most commonly found right at the beginning of the paper. One paper
in my corpus began with a series of questions:

(15) How do students solve mathematical problems that contain an equivalent
      mathematical structure, but variance in the narrative situation presented in
      them? Are mathematical problems that appear as part of a concrete situation
      easier to solve than those that appear in an abstract situation? What is the
      effect of the different learning conditions on the development of the ability to
      solve problems presented in different situations? The following study touches
      on these questions with a focus on the problems discussed in the linear func-
      tion graph.   (660)

Formulations of this kind position the goals of the writer right at the beginning
of the text and present the entire paper as an attempt to achieve these goals, i.e. to

answer the questions. In other papers, questions such as these appeared, although not right at the beginning but rather in the body of the Introduction, and occasionally at its end, as an element that leads into the Method section.

What is the dialogic effect of questions such as these? Do they involve any interaction between the author and reader?

Thompson and Thetela (1995: 124) present an intriguing idea. In their view, the strategy of asking questions that the author answers himself later on expresses an interaction not between the author and reader, but between two states of the author himself: between the author in the "past," at the time when he began his research (the stage at which he is asking) and the author in the "present," at the time he is writing the paper (at the stage when he can provide answers to the question). This interaction between the states of the author can also be expressed in the use of various grammatical tenses – future/present at the stage of the question as compared to present/past at the answer stage. Nevertheless, the rhetorical effect of these questions is related to the reader sharing in the cognitive process that underlies the text. The reader is invited to participate in this process, and his interest is piqued in wake of the questions, as well as when he understands and receives the answers provided by the author.

The questions can appear as subheadings and thereby serve as markers of structure directed at the reader. In one paper, I found a series of subheadings in the Findings section. In fact, this appears to be a breakdown of the research questions, which appears interspersed, and each question precedes the findings that can answer it:

(16) Presence: Who appears more?   (472)
      Location: Where do men live and work?   (474)
      How does one convince men and women?   (475)
      How are men and women presented?   (477)

These questions essentially present a dialogue with the reader, as Webber notes, "Questions create anticipation, arouse interest, challenge the reader into thinking about the topic of the text, and have a direct appeal in bringing the second person into a kind of dialogue with the writer" (Webber 1994: 266). These are questions that are answered in the text, and in this sense are real questions. On the other hand, the next category that I will present contains real questions that are part of a dialogue with the discourse community as a whole.

(b) *Open questions directed to the disciplinary community.* "Real questions" positioned towards the end of the paper often represent unresolved problems that may serve as the subject of future research. Questions such as these are not generally answered in the paper itself. As Adams Smith (1987: 19–20) pointed out, "it is common for a piece of research to answer the question it has set out to clarify, while at the same time it raises other questions to be accounted for in the course of further investigation." Here are two examples from my corpus:

(17)   Can the structuring of moral communities [...], even if on a local level, which include an executive, teaching staff, students and parents, help teachers when contending with ethical issues? And perhaps it is important to create the teachers' identification with their own community, as one that has certain qualities and that champions certain ideals? These possibilities need to be explored and examined.   (463)

(18)   Is it possible to generalize from the findings of an Israeli sample regarding the effects of military service in general? It is difficult to answer this question without carrying out a comparative study of other military systems. Nevertheless, it should be noted that any comparison with other armies would have to contend with the difficulties that stem from the nature of IDF service [...].   (612)

The status of the author is not weakened in wake of presenting questions that he is unable to answer, because he answered other questions, those that were at the focus of his discussion. On the contrary: the formulation of new questions in wake of the study enables the author to establish his own niche within the entire scientific endeavor along with his standing as a valuable member of the disciplinary community. Questions, problems or unexplained phenomena are the essence of scientific work (Swales 1990: 140). The importance of the current study increases due to the fact that its outcomes raise new questions that awaken the need for additional research. It is through this that the cumulative and continuing nature of the scientific endeavor is consolidated, that to which the author had the privilege of adding yet another tier (see also Montgomery 1996: 38). In doing so, the paper manages to establish additional territory, with an eye to the future (see Section 3.4 above).

Questions such as these, that turn to the future, to the additional territory, resemble in their function utterances that include recommendations and proposals for future research, which are often found towards the end of papers, for example:

(19)   Furthermore, *it is suggested* that this question be explored for other ages (childhood) and in different populations.   (436)

(20)   In further studies, *it should be explored* as to whether this pattern is characteristic of academic professions [...].   (277)

(21)   [...] *It is recommended* to study the verbal interaction in a small group [...].
                                                                                      (678)

In Hebrew, recommendations and proposals of this kind are formulated impersonally, mainly by means of the syntactical construction: *modal/evaluative operator + infinitive*, for example:

> *mutza livchon* (lit. "suggested to explore")
> *yesh livdok* (lit. "there is to explore")
> *mumlatz lachkor* (lit. "recommended to study")

In Hebrew, a language that is tolerant of subjectless constructions, this construction is a powerful tool for designing an impersonal tone, especially in academic discourse (Livnat 2010b). The impersonal tone implies that what the author is proposing could be carried out by him, just as it could be carried out by any other member of the scientific community. Consequently, these utterances are important rhetorically in two ways:

i.   They are related to the establishment and occupation of a niche. By proposing a way to develop his ideas in the future, the author gives his research a clear place within the coherent continuum of the entire scientific endeavor.
ii.   As a result of this, his proposal also contributes to strengthening and stabilizing the status of the researcher as a member in good standing of the disciplinary community, one who not only carries out valuable scientific work, but also proposes to other members of the community how to use it in the future.

Lindeberg (2004) found in papers in English that in many cases suggestions for further research are marked by positive evaluation, which is a further way of promoting the new claims in the paper (p. 160).

It is true that utterances of this kind, exactly like questions that are directed at future research, often merely hint at the next paper that the author plans to write. Hyland expresses this by means of a tongue-in-cheek rhetorical question of his own: "Should we see this as a subtle staking out of a claim for priority and a promotion for a forthcoming paper?" (553). However, the most important aspect as I see it is that these are real, open questions that remain unanswered in the paper itself. The reader is not expected to answer them either because the answer has not yet been found. The author presents these questions for the scrutiny of the disciplinary community, and they consequently reflect the dialogue that he is holding, not necessarily with the reader, but with the discourse community as a whole.

To sum up, it appears that direct questions – both real and non-genuine – fulfill a number of rhetorical functions in academic papers. They create a dialogue with the reader and bring him into the logical and cognitive processes that underlie it. They draw the reader's attention to certain points in the argument and propel him in the direction the author is aiming at. They help to organize the discourse, make a smooth transition among the paper's various parts and highlight the structure of the argument for the reader. When they appear in the title of papers, they attract the potential reader's attention, stimulate and challenge him.

However, real questions that appear at the end of a paper and are directed at future research are addressed to the scientific community and its shared project, offer it new research directions and thereby establish the status and importance of the new knowledge offered by the paper.

4.6 Scientific dialogicity: A combined model

In the current chapter, we have looked at four textual components that are involved in the shaping of the dialogic nature of the discourse of academic papers. As Table 1 above shows, there are many other components that I have not discussed, but these four will suffice to exemplify and justify the combined model that I proposed in Section 4.1 above.

In order to complete this model, I will add yet another important component, one that has been mentioned repeatedly in this study – the use of genre conventions. The very fact of the acceptance and use of genre-based conventions shared by the discourse community creates a dialogue with that community, which is an important component of the constitutive dialogicity.

The model that I propose integrates the theoretical approaches suggested above in Chapter 2 with the analyses that I have suggested in the current chapter. It clearly distinguishes between the act of addressing the reader in reading situations and calling for reader participation, while striving for *responsive understanding* (Bakhtin 1986), on the one hand, and a more general address of the discourse community, on the other. This distinction is illustrated in the model in a number of ways. First, it distinguishes between two types of *we*: the inclusive *we* that aims to include the reader, and the communal *we*, which is directed at the research community as a whole. Second, it distinguishes between the two different functions of questions: questions that are related to bringing the reader into the text and sharing the reading with him, and questions that are addressed to the research community, that open the path to new research and point to additional issues that may concern researchers and others in the future. Concession structures can also serve in these two separate functions: some are aimed at noting reservations that the reader himself might entertain while reading, while others maintain a dialogue with the entire discourse community by referring to shared knowledge and norms.

Table 6. A combined model of scientific dialogicity

Dialogue with				
	Reader		Concession (reference to an assumed claim or reservation)	Manifest intertextuality
			Inclusive we for reader's participation	
			Non-genuine questions	
	Discourse community	Specific member	Citations	
			Concession (reference to a specific claim)	
		Non-specific member	Non-attributed Citations	
			Concession (reference to a non-attributed specific claim)	
		The community as a whole	Concession (reference to shared knowledge and norms)	
			Communal we	
			Open questions directed to the scientific project	
			Use of genre conventions	Constitutive intertextuality

In addition, the model differentiates between different "voices" (Thompson 1996) – both specific and non-specific. These terms by Thompson make it possible to distinguish between the mention of a specific member of a discourse community and reference to an argument made by a nonspecific member, a less common albeit existing phenomenon. Similarly, it is also important to distinguish between the mention of individual members in the community who may hold specific but not necessarily uniform views and referring in general to the members of the community and what they have in common.

The model also makes it possible to see the place of constitutive intertextuality within the entire system of scientific dialogicity. Unlike the manifest dialogicity, which is expressed on the surface, constitutive dialogicity is deeply embedded in the text and consequently requires a far more attentive reading.

This model will serve me in the coming chapter to analyze entire academic papers and present a number of distinct types of scientific dialogicity. Following the intertextual analysis in Chapter 5, in Chapter 6, I will present a complete list of the linguistic components that design scientific dialogicity.

CHAPTER 5

Scientific dialogicity in action

5.1 Introduction

The notion of *dialogicity* may take on a more transparent meaning when an actual scientific dispute is being explored. In order to take a closer look at the dialogic dimension of academic discourse, in the present chapter I will propose an analysis of a number of complete articles that focus on the same controversial topic and hold a dialogue with one another. This will enable the reader to gain an in-depth understanding of the dialogic processes that the writers engage in, based on a certain familiarity with their research questions. Through this analysis, various patterns of scientific dialogicity will be described, according to the different levels of directness and confrontation they engage in with others.

As I noted in Chapter 3 above, most academic papers present new data or ideas that the researcher wishes to put forward in his or her new paper. The contribution of these papers is closely related to the body of knowledge of that discipline. The writer's goal is to introduce new knowledge and to make sure that it will find its place within the discipline's shared reservoir of knowledge. In other cases, the author possesses a status that enables him or her to present material that is not new and may not necessarily be the result of his or her own work.

On the other hand, in some cases, the author inserts himself into a debate, not because he seeks to advance his new claims, but rather due to an oppositional stance. He may be opposed to a previous paper written by a specific scholar or to the ideas represented by an entire school of researchers. In a case such as this, an entire ping-pong of responses and counter-responses may develop.

Each of the situations described here involves a different type of behavior vis-à-vis the discourse community, specific members of the community and the reader. A picture of scientific dialogicity would not be complete without a description of all the various forms it can take.

The manner in which the academic paper, through its structure and language, reflects academic conflicts has been the subject of considerable interest on the part of numerous researchers. Some explored the degree of criticism expressed in academic papers in comparison to other genres. Kourilova (1996), Motta-Roth (1998) and Hyland (2000), among others, have shown the differences between journal articles and book reviews or referees' comments on manuscripts submitted for publication from this perspective. These latter genres are evaluative by definition and are therefore expected to involve a higher degree of personal conflict. In the research article itself, the abstract is the part where criticism or negative evaluation can be expected to be manifest (Stotesbury 2006).

Some researchers investigated the criticism expressed in papers or abstracts only, using the model developed by Swales to create a research space (Ahmad 1997; Burgess 2002). As described in Section 3.1 above, one of the steps in creating a research space is to establish a niche. In order to secure his niche, the researcher must point to a gap or deficiency in the existing knowledge. This action by nature forces him to confront the world of research, and consequently, the assumption is that the appearance of this step in the text promotes a confrontation, while its absence makes it possible to avoid one.

Academic criticism has also been analyzed from a diachronic perspective (Salager-Meyer 1998b, 1999). Many researchers have examined this question from a cross-cultural perspective by comparing papers written in different languages (Taylor and Chen 1991; Bloch and Chi 1995; Duszak 1994, 1997; Čmejrková 1996; Salager-Meyer 2001; Salager-Meyer et al. 2003; Martín-Martín 2005; Giannoni 2005).

Martín-Martín (2005:86–88) proposed a taxonomy of strategies to convey academic conflict based on some of these previous works. His model describes three dimensions on which a criticism can be expressed:

a. *Personal* and *Impersonal*. In the former strategy of expressing criticism, the name of the researcher who is the target of the criticism is explicitly mentioned, whereas in the latter, the criticism is directed towards a particular position or at the discourse community as a whole.
b. *Direct* and *Indirect*. In the former, there is a categorical criticism, whereas in the latter the criticism is mitigated by means of hedges.
c. *Writer mediated, non-mediated* or *reported*. In the former, the writer is explicitly present in a critical speech act by the use of the first-person pronoun. At the opposite end of the continuum, the author merely reports criticism leveled by another author.

The criticism expressed in a paper can range along any of these dimensions. In general terms, it may be claimed that the more personal, direct and writer-mediated the criticism is, the more confrontational the paper will be.

This model relates to a number of elements that were discussed in the previous chapter. One is the use of the first-person pronoun (in the singular or exclusive-plural), whose appearance Martín-Martín assumes can increase the confrontational nature of the text. The second is an explicit quote from the work of another author who is mentioned by name, which makes the criticism towards him more personal. Martín-Martín did not relate to the different patterns of citations (integral or non-integral) and the impact that each pattern can have on the extent of the other's salience. He uses this model to compare the writing of abstracts in English and Spanish, although he did not try to present different models of papers based on how confrontational they are, something that I will try to do in the current chapter.

An important contribution in this direction was made by Susan Hunston in a number of her corpus-based studies. Hunston describes the typical dialogic situation of the academic paper in general terms thus:

> Most academic research articles construe opposition with other researchers, even if only in passing. To test and possibly refute the theories of others is one of the accepted justifications for writing such an article. Such opposition is normally construed within the constraints of research article conventions, including principles of politeness.
> (p. 1)

Myers discusses politeness in scientific discourse at length (1989), and explores the academic paper from the perspective of politeness theory (Brown and Levinson 1987). Myers claims that in many cases of scientific communication, certain face-threatening acts are avoidable, and must be redressed with various politeness devices. Every scientific report states a claim that is to be taken as the article's contribution to knowledge. Most reports, in stating a claim, deny or supersede the claims of others. Myers thus described the role of various linguistic components as positive and negative politeness strategies. One might say that politeness itself is an expression of the author's dialogue with the discourse community and the calculated and cautious manner in which he positions himself within a relationship with other members of the community. One expression of politeness in scientific papers is that the innovation in them is foregrounded by means of the new data provided, and thus responsibility for the opposition is assigned to the results. "This is polite," Hunston argues, "in that it construes opposition between research outcomes as opposed to construing an argument between researchers, but it is also persuasive, in that the argument is presented in a way that appears to be 'objective' rather than personal" (2005: 2).

In this way, it is consequently possible to describe the dialogic nature of the academic paper, which is necessary by virtue of the very role of this genre in the discourse community. The assumption is that most papers reflect a moderate and polite dialogicity of this kind. On this background, papers of a different dialogic nature stand out. And this is what Hunston (2005) writes on papers of this kind:

There is, however, a type of research paper which obeys a very different set of conventions. These articles declare their purpose to be specifically to counter the opinions expressed in previously-published articles. They do not present new research but engage in an argument that is more overt and personal than that found in typical research articles. (p. 2)

These 'conflict articles'

[A]re by definition responses to what I will call initiating articles: initiation and response comprise a 'conflict exchange.' Any article might retrospectively be cast as the initial element in a conflict exchange, although articles so identified do themselves often take a contentious stance and might be said to be deliberately controversial. Unlike exchanges in spoken dialogue, the conflict exchange has no defined end; any number of articles may respond to each other, though in practice the maximum number of moves seems to be three. (ibid.)

The difference in principle between regular papers and 'conflict articles' lies then in their declared aim, which is usually evident from their title. Hunston used corpus analysis to characterize the unique linguistic and rhetorical aspects of the conflict article. In Hunston's (2005) study, 13 conflict articles from *Applied Linguistics* were compared with a 2 million-word corpus of general applied linguistics articles from a range of journals. She shows how conflict articles construe conflict and at the same time construe consensus between writer and reader (p. 13). The main linguistic difference that Hunston found between the conflict and regular articles is expressed in elements that are related to what she calls *evaluation*.

For Hunston, evaluation is "the broad cover term for the expression of the speaker or writer's attitude or stance towards, viewpoint on, or feelings about the entities or propositions that he or she is talking about" (Thompson and Hunston 2001:5). According to Hunston (1994), evaluation may be viewed as being of three kinds: (1) Factual *status* – certain-uncertain (see Section 3.3 above); (2) *relevance* – important-unimportant; (3) *value* – good-bad.

The term 'value' refers to "that aspect of evaluation in which the values 'good' and 'bad' are ascribed to an entity. [...] What counts as good or bad depends on the bases for value-assignment available to a given discourse community" (Hunston 2005:4).

"Evaluation both exploits and construes a value system belonging to the relevant community; members of the community recognize that certain qualities have a positive value while others have a negative one" (p. 4). One feature that may distinguish conflict articles from regular ones is that conflict articles draw on, and make very visible, a specific value system, often criticizing previous researchers rather than previous research in a way that is unusual in academic discourse (p. 12).

> Although the same bases are used in 'ordinary' research articles, they are less frequently given negative value than in these articles. Arguments are more usually evaluated negatively by being shown not to fit with experimental results, or to be 'unreasonable' in that they do not accord with current thinking in a particular discipline (Hunston 1989, 1993). Even when negative value is accorded, it is more likely to be attached to the outcomes of research [...] than to the researchers themselves. (p. 5)

As we shall see in the current chapter, what Hunston calls a "conflict pattern" can in fact be expressed in a number of different forms. One possibility, one that Hunston notes, is that the author specifically relates to a paper written by another author, so that the combination of the two creates what may be called a 'conflict exchange'. Hunston chose to analyze only papers of this kind. The first author may respond, creating a pattern that I call the 'ping-pong pattern', and which I will demonstrate in Section 5.6. On the other hand, some articles having a confrontational nature do not conform to this description. They do not define themselves in their title as "A response to...". However, their contentious aim immediately becomes clear upon reading them. They do not necessarily confront a particular author of a particular article, but may in fact relate to the ideas of a specific school of researchers and a large number of publications, as I will demonstrate in Section 5.4 and 5.5.

To the written patterns of the dialogicity, I have added in Section 5.7 a description of a face-to-face confrontation focusing on the same questions, which were recorded and published in a semi-popular journal. The various patterns will be presented based on how direct their dialogicity is, from the most indirect to the direct. The face-to-face interaction is the most confrontational pattern, and consequently appears at the end of the chapter. Although my research generally deals with the written text rather than spoken interaction, I have added the last section for the sake of comparison and in order to enrich the general picture of debate with yet another aspect.

5.2 The United Monarchy: A question from the past

I have chosen for analysis a number of academic papers that are part of a scientific debate that has been held in the last few decades in Israel and beyond in regard to a certain aspect of ancient Jewish history. Archaeologists, historians and biblical scholars – members of different disciplines in the humanities – are involved in this debate, which is an ongoing polemic held on the pages of scientific journals as well as in the Israeli and foreign media, since it is a subject highly charged with political and ideological significance. To one degree or another, it touches on the

Jewish and Israeli identity of many people. It is loaded with emotion and often involves blunt language, derogatory epithets, *ad hominem* accusations and an attack on the ethos of the opposing scientists. The background of the debate will be described below.

According to the Bible, as matters are presented in the books of Samuel and Kings, the city of Jerusalem was at the center of the reigns of two of the greatest and most important monarchs in the history of the Jewish nation – King David and King Solomon. In the first seven years of his reign, David ruled over only the tribe of Judah, with his capital in Hebron. After seven years, David succeeded in uniting the 12 tribes of Israel. He conquered Jerusalem from the Jebusites and established it as his capital, from which he reigned for another 33 years. He set himself up in an existing citadel in the city known as the "Citadel of Zion" and renamed Jerusalem the "City of David." The scriptures tell us that during David's reign, Hiram King of Hazor built a palace in Jerusalem. David's reign was filled with wars and conquests, in the wake of which his kingdom expanded considerably to become an empire stretching from the Euphrates to Egypt.

David's son and heir Solomon ruled for 40 years, and it was during his reign that the United Monarchy achieved its greatest influence. According to the Bible, Solomon managed to avoid fighting wars during all the years of his reign; he married the daughter of the king of Egypt, maintained extensive political and commercial ties with distant lands and levied high taxes in all the inhabitants of the region under his rule. Solomon was also responsible for a number of ambitious construction projects: He built himself a magnificent palace in Jerusalem in addition to the First Temple, to whose highly detailed description an entire chapter of the bible is dedicated (I Kings 6). The Bible provides us with a comprehensive list of Solomon's considerable wealth and the cities under his control, and chronicles the intricate system of tax collection throughout the kingdom. Immediately after Solomon's death, the kingdom split in two for 200 years, never again to be reunited. According to the Bible, the southern kingdom of Judah, whose capital was Jerusalem, remained the more important of the two parts, and the monotheistic religion continued to flourish in it, in the glorious temple that Solomon built. The northern kingdom of Israel, whose capital was in the city of Samaria, however, is presented in the biblical text as being of lesser importance and as a kingdom of inveterate sinners.

The reigns of David and Solomon, which are customarily dated to the tenth century BCE, are consequently described in the Bible as the most glorious period in the history of the Jewish people, as an entity that enjoyed political independence, economic might and regional influence. This was seemingly the only period when the entire nation was united under a single central government, with its capital in Jerusalem, which was also recognized as a site sacred to the monotheistic religion. This explains the enormous significance and importance of this period in the eyes of both Israelis and Jews today too.

For many generations, the biblical text was considered by Jews to be a quasi-historical account containing a significant kernel of truth. After the establishment of the modern State of Israel, this approach dominated academic research in the universities and educational system as part of the process of creating a sense of Jewish-Israeli nationalism. Every child that grew up and was educated in modern Israel knew of David and Solomon as historical figures that lived in the land of Israel, powerful kings who built the city of Jerusalem, and also human beings with all their human frailties and passions, as the Bible describes them so well. A large proportion of Israelis today appear to assume, either consciously or unconsciously, that these kings are their forebears.

However, in recent decades, reservations have been increasingly voiced in the academic world regarding the extent to which this biblical story (along with other parts of the Bible) conforms to actual historical truth. In an article that appeared in *Haaretz* in 1999, Ze'ev Herzog, an archaeologist from Tel Aviv University, maintained that the biblical story regarding the great unified kingdom of David and Solomon is no more than an imaginary historiosophic tale. His article, "The Bible: No findings on the ground,"[5] brought these questions to the attention of the general public and provoked considerable interest and controversy among the Israeli public notwithstanding the fact that the question of the consistency of the biblical stories with archaeological findings has been grist for the mill of academic research for the past one hundred years. Biblical researchers, historians and archaeologists differ on this matter in many aspects, including some that are related to the historical reality that prevailed in the tenth century BCE: Who was the "Israeli nation" described in the books of Samuel and Kings? Where did it come from? What was its religious faith? How big was it or could it have been? Did the "United Monarchy" of David and Solomon even exist in the tenth century BCE? And if so, how big and powerful was it really? Are the biblical kings David and Solomon historical figures or merely the stuff of legends?

Many, albeit not all scholars today believe that the biblical texts that describe the reigns of David and Solomon were not written close to the time when the events described in them occurred, although they may be based on older written texts. Many claim that a class of scribes able to carry out a task as complex as writing the Bible did not yet exist during this period. Some researchers believe that the texts were composed in the seventh century BCE, about 300 years after

---

5.   In the English version of the newspaper the title was: "Deconstructing the Walls of Jericho."

the events that they describe, and that they reflect the dominant ideology during this later period. The most extreme are those who belong to what is known as the "minimalist" school, who claim that the books of the Bible were written during the Persian-Hellenist period and that they have no historical value whatsoever. In their view, the Bible is a later invention aimed at creating an imagined "past," a false "collective memory," for a group of people lacking any shared primary national history.

In the year 2003, archaeologist Israel Finkelstein, the head of the Archaeological Institute of Tel Aviv University at the time, gave an interview to *Haaretz* following the publication of his book (written together with Neil Asher Silberman), *The Bible Unearthed: Archaeology's New Vision of Ancient Israel and the Origin of Its Sacred Texts*, which was published in a number of languages and caused quite an uproar both in Israel and abroad. Finkelstein maintained that Jerusalem in the tenth century BCE, the period to which the United Monarchy at the height of its glory is dated, was no more than one of many typical mountain villages having no particular regional importance. At most, it may have contained a modest temple that served the religious needs of a small tribe. The Biblical description of the United Monarchy, in his view, is a religious-political manifest written many years later, in the seventh century BCE. During that time, which came in the wake of the destruction of the kingdom of Israel, the southern kingdom of Judah was flourishing. This was the time when it consolidated its own self-determination along with its monotheistic approach: one state, one capital, one temple, one king, one God. According to this approach, the account of the United Monarchy of David and Solomon was a legend cultivated to serve this ideology.

These questions currently lie at the center of an acrimonious debate among archaeologists, historians and Biblical scholars, both Israelis and others. The dispute is often described as being part of a genuine revolution that the world of Israeli archaeology is undergoing. This involves a transition from archaeological research that draws on biblical texts that it seeks to corroborate in the field, to an alternative paradigm. This revolution began with a crisis that was precipitated when it was discovered that the archaeological findings do not actually support the biblical story. Gradually, doubts regarding the historical value of the stories of the patriarchs – Abraham, Isaac and Jacob – the exodus from Egypt and Joshua's conquest of the land of Israel began to arise. Today, one would be hard pressed to find a reputable archaeologist who accepts the biblical text at face value as a historical testimony in all its details on these matters, even if it is viewed to one extent or another as preserving ancient traditions in regard to events that accumulated over time in historical memory.

On the other hand, from the ninth century BCE (after the United Monarchy ruptured), the biblical stories are supported by numerous external sources. The scientific crisis then appears to focus on events that occurred in the tenth century BCE. As noted, this crisis is especially virulent because of its potential ideological, theological and political implications. This is the background for the polemic that lies at the center of the debate that is investigated in the present chapter.

The perspective that I will be presenting here is linguistic-rhetorical: What does a scientific polemic look like? What are the types of arguments that are raised in it? What are the linguistic components that reflect the dialogic nature of the text? How is the dialogue carried out with those that hold opposing views, with the research community as a whole, with the disciplinary conventions, with the reader and the general public? How are the "facts" and "proofs" presented from a linguistic perspective? These are the questions that I will try to answer here.

It is in the nature of the academic text that it is always based on previous, existing texts. There is never any possibility of starting "from scratch." Every piece of writing that presents a new claim quotes from previous texts, and the new claim is new in relation to them; going back in wake of the cited texts is endless. Consequently, some starting point has to be arbitrarily set with the polemic investigated from that point forward. The papers that I have chosen to analyze were authored from the mid-nineties on and earlier texts are only referenced in them. The texts that were chosen cover only a small part of everything that has been written on this issue. It is a given that the issues that are being raised here, along with their respective disciplines, are both complex and multifaceted. Consequently, this brief survey cannot fully represent their complexity and nor is this the goal of this discussion. To that end, the reader is directed to the papers themselves and the references that appear in them. From the entire available selection, I have tried to choose significant items that are cited by others and that relate to one another, which together will give readers a reasonable picture of the debate and enable them to understand the main arguments involved. The papers have been selected to reflect different types of scientific dialogicity, as will be described below. The full list of articles appears in Appendix B.

Looking at the various dimensions of dialogicity, as they were described in Chapter 4 above, the rhetorical analysis of this debate will demonstrate the ways in which the scientific discourse maintains dialogicity. In addition, the analysis will pose questions related to the modes of scientific arguments, the types of relevant proof for the different disciplines, the way in which "facticity" is presented, the status of "objectivity" in the humanities, the ethos of the scientists, the place that emotion and ideology occupy in science and more.

5.3 The classic pattern
(Steiner 1998)

The paper that I will be analyzing in this section (Paper 1 in Appendix B) represents the classic, non-confrontational dialogicity. The presentation of the paper in this section has two purposes: First, its dialogic character will serve as a kind of "control group" to which we can compare the confrontational models. Second, the content, types of arguments and proofs that it presents will serve as background to help us understand the specific issues under discussion in the other papers too.

Margreet Steiner is an archeologist in Rijks University, Leiden in the Netherlands. The aim of her paper is to present and analyze the archaeological findings from Jerusalem in the Bronze and Iron Ages, and through them to provide a picture of the city in these periods.

Her bibliographical list includes 122 items, including six written by the author. The length of the list is indicative of the paper's intention to be conclusive and all-embracing. The paper itself contains 188 references, of which 162 (approximately 86 percent) are non-integral citations (see Section 4.2.1 above). These are in part sources that support her thesis and describe archaeological findings that the author will base herself on when she presents the archaeological picture. Other sources contain interpretations of the findings. She mentions some of them in order to challenge them and refute the interpretations that they offer.

The introduction to the paper describes the history of the excavations in Jerusalem and the methodological limitations of each of the researchers involved in the excavations. This introduction creates Steiner's research space, not in the sense of a deficiency that the paper intends to fill, but in that it sketches out the general background and main sources for the findings, as a preparation for the detailed description that the paper will provide.

The only citation of an entire paragraph appears right at the beginning of the introduction; it is a citation from the British archaeologist Kathleen M. Kenyon – one of the most important archeologists to excavate Jerusalem – that explains why she decided to dig in Jerusalem. It would appear that the purpose of highlighting this particular citation, besides the important place that it accords Kenyon, is to distinguish between her and 'biblical archaeologists,' who conduct archaeological research as part of Bible studies, a prevalent trend in the archaeology of Eretz Israel from the late 19th century up to the past few decades. The inclusive pronoun *our* that appears in Example (1) in the context of Kenyon's excavations is interpreted in the broadest way possible to refer to all those who aspire to reach the truth about the history of the world:

(1) For her [= Kenyon], the archaeology of Palestine was a branch of general world archaeology, and its task, therefore, was to enrich *our* knowledge of the ancient world by systematic research and strict application of scientific methods.   (p. 144)

This sentence also notes the principal conventions of the discourse community – the importance of systematic research and the strict application of scientific methods. These are values that the members of the research community will immediately identify as being positive, creating a research environment having an objective nature and that is free of ideology and emotions that could bias the research. The emphasis on this direction is also evident in the sentence that introduces the next chapter of her paper, in particular the word *solely*:

(2) This and the subsequent sections will survey the development of the ancient town of Jerusalem based *solely* on the archaeological remains discovered during recent (and older) excavations.   (p. 145)

When Steiner notes here that her work will be based solely on archaeology, she is already hinting at a lack of consistency between the archaeological findings and other sources, to which she will relate in the coming chapters. The chapters that follow describe the development of the city in chronological terms, from the middle Bronze Age (c. 1800 BCE) until the destruction of the city by the Babylonians in the Iron Age II (587 BCE). Steiner's general claim is that the archaeological findings do not provide proof of continuity of settlement in Jerusalem and that there are periods for which there are no actual findings. This claim, which is the conclusion of the paper, is associated with a number of confrontations with researchers that hold opposing views, who base their views on sources other than findings from the excavations in Jerusalem.

One such source is the name *Rushalimmu* that was inscribed on broken pottery sherds discovered in Egypt dated to the 19th or 18th centuries BCE. The sherds contained the names of the enemies of Egypt, and were probably broken as part of a ritual act of magic. The fact that the city is noted as an enemy of Egypt, which was a hugely powerful empire at this time, seems to indicate that Jerusalem at this time was a significant city. At this stage, Steiner offers a paragraph that may be a reflection of a dialogue that could be interpreted as her own internal dialogue or of an external debate actually taking place:

(3) However, the use of this name by itself cannot provide 'proof' that Jerusalem was an important city at that time. The name does not necessarily specify a town. It could as easily indicate a region or a tribe. Historically, a small provincial town such as the excavations show Jerusalem to have been at this time would have been unlikely to bother the mighty pharaoh of Egypt.

In the terms used by Martín-Martín (2005, see Section 5.1 above), Steiner's criticism here is impersonal and reported, and can be viewed as a means of diminishing the criticism and its adversarial nature. At the end of the paragraph, it appears that Steiner has a position on the subject under discussion, although it is not stated explicitly in the text.

By contrast, immediately afterwards, Steiner mentions Finkelstein's article (1992) and expressly repudiates his claim regarding the status of Jerusalem during this period based on the archaeological findings. In his 1992' paper, Finkelstein argues that in the Middle Bronze Age, a development into larger political entities would have taken place in the hill country, with Shechem and Jerusalem as the center of larger units. The rejection of his claim is supported by explicit negative expressions:

> (4)   According to the archaeological remains, however, Jerusalem *cannot* have played this role. As far as is known […] *no* architectural remains and *no* pottery from the second half of the Middle Bronze Age was found in any of the excavations […]. (p. 148)

Fløttum et al. (2006: 244–245) consider these explicit negative expressions to have polemic value and the potential to refute the claim. For Hyland (2009: 38), negation is a forceful and dialogistic means of engaging with others' views and disputing alternatives, a resource for introducing an alternative position into the dialogue in order to reject it.

Going back to Steiner's paper, another important external source for understanding the history of Jerusalem is from the Late Bronze Age (1550–1200 BCE). In el-Amarna in Egypt, six letters written by the scribe of Abdi Khiba, prince of Urusalim, dated to the second half of the 14th century were found. Steiner presents the position of scholars who tried to deduce the nature, dimension and status of Jerusalem during this period from the content of these writings. The paragraph mentions three specific scholars, although their names are noted in parentheses in a non-integral reference, and the sentences formulated in the passive voice:

> (5)   Based on the content of these letters, most studies describe Jerusalem as a large town protected by sturdy walls. It is *assumed to be* the center of a city state (Bunimovitz 1995: 326), the seat of the ruler of a dimorphic chiefdom (Finkelstein 1993: 122), or the commercial centre for the immediate region (Thompson 1992: 332). This function is then *supposed to be* a continuation of the site's position during the Middle Bronze Age. (p. 149)

Concerning the passive voice, Myers (1989) suggested that passive formulations in the scientific paper serve as a form of politeness, and as such should be viewed as a means to diminish the confrontation between the author of the paper and the cited scholars, who are colleagues and members of the same discourse community.

The reason for Steiner's opposition to the noted claims is that the picture that this source presents is at odds with the archaeological findings in Jerusalem itself. Steiner uses a self-citation here of a study in which she discussed this contradiction:

(6) When *one* concentrates on the archaeological remains found in Jerusalem from the Late Bronze Age, a different picture arises. It has already *been argued* that the large terrace system [...] cannot be dated to the Amarna period (M. Steiner 1994). This leaves *us* with very little material originating in the Late Bronze Age [...].   (p. 149)

Despite the fact that what is involved here is a claim for which the author takes full responsibility, from a linguistic point of view, her presence in these statements is not prominent. Three linguistic means are used here to diminish her presence: In the first sentence, it is the indefinite generic pronoun 'one', which means "people in general" (Quirk et al. 1972:222), has no "personal overtones" (Biber et al. 1999:331), and thus indicates an "indefinite author" (Fløttum et al. 2006:110–112). The second sentence is written in the passive voice, and the pronoun *us* in the third sentence is inclusive and includes within it the entire community of researchers. Her criticism is then non-mediated, in Martín-Martín's terms (2005), and consequently mitigated.

In the unequivocal wording of the conclusion that immediately follows, on the other hand, we find her only use of the first-person singular pronoun in the entire article:

(7) *To me*, the conclusion seems inevitable that no 'city' existed in Jerusalem during the period of the Amarna letters, even though *most of my colleagues* do not agree with this.   (p. 149)

This linguistic difference reflects the author's desire to underscore her conclusion, which runs counter to that of others (*most of my colleagues*). We will once again encounter this position on the part of Steiner in regard to the Amarna letters when we analyze the article by Na'aman in the next section.

Steiner continues to challenge the conclusions of others when she sums up and presents her conclusions in regard to later periods in the history of Jerusalem too:

(8) As there is no evidence for any occupation in Jerusalem from the seventeenth century BCE onwards (and for the fourteenth century *we* have only the

> Amarna letters), *it seems hardly possible* to picture Jerusalem as the centre
> of a city-state [...] (contra Finkelstein, Bunimovitz, Thompson – see above).
> (p. 150)

The adversarial atmosphere that the author creates in this text is restrained and toned down. This tempering effect is created by the mention of the names of other researchers in parentheses, in a cluster of references, without specifically mentioning any one of them, and with a back reference. Her criticism is felt to be indirect, as put by Martín-Martín, due to the hedge phrase *it hardly seems possible.* The pronoun *we* also creates more of a feeling of cooperation than confrontation, because it is interpreted to be referring to the entire community of researchers of which the author is part. Another way to understand it is as addressing the readers and sharing the process of argumentation with them, a strategy by means of which the author leads the readers through the argument to the conclusion that is meant to be drawn from the data that is presented.

As Steiner's description progresses, she continues to hint at disagreements in the research. The chapter that deals with the Iron Age (1200–600 BCE) refers to a number of points of disagreement. First, the absence of any archaeological findings from the years that preceded the establishment of the "United Monarchy" (1200–1000 BCE) contradicts the Biblical texts that relate that David conquered the city, captured the citadel and settled in it. Here Steiner is particularly unclear in a dialogue that she creates with other researchers who used the Biblical text as historical evidence:

> (9)   Therefore, the traditional picture of Jerusalem in this period presented *in
>   most books* is that it was a small, well-fortified town inhabited by Jebusites,
>   the centre of an independent city-state. Later on, this town was taken by King
>   David and transformed into his capital (see, e.g., Mendelhall 1989). (p. 150)

The expression *in most books* refers to other researchers in very general terms, whereas only one is mentioned specifically by name in a non-integral reference that does not emphasize his presence. Here too then, the challenging of those holding other views is not underscored and the criticism remains quite impersonal.

In this section, the dialogicity with the reader is more greatly emphasized, and is designed by a number of linguistic means. In the context of presenting her argument, Steiner uses the indefinite *one*, obligation modals that target the reader (Hyland 2001a) and questions, which are worded as if the reader could also have raised them at this point:

> (10)   *One has to assume* that the important buildings connected with this system
>   were constructed on top of the hill.   (p. 150)

(11)   *One must ask* why such an enormous task was undertaken.   (p. 150)

(12)   What kind of building adorned the top of the hill?   (p. 151)

Both questions (in (11) and (12)) engage the reader only to a limited extent, because the answer to the questions is provided immediately afterwards. The question in (11) is also worded as an indirect question and both are related to the organization of the discourse and address the reader in that they mark out the transition to the new subject, which is worded in the form of a question.

The use of the first-person plural pronoun is another linguistic means aimed at engaging the reader:

(13)   Thus, instead of a town *we* have a small fortified stronghold that required a
       great deal of effort to build.   (p. 151)

An interesting appearance of the first-person plural is in the following paragraph:

(14)   According to Thompson, Jerusalem functioned throughout the Late Bronze
       Age and Early Iron Age as a centre of commerce and trade for the small agri-
       cultural settlements nearby. [...] Only in the eighth and seventh centuries was
       the site transformed into the capital of a regional state [...] (1992: 333). This
       would also be the earliest period in which *we* could expect to find the forma-
       tion of scribal schools.   (p. 151)

The paragraph is devoted to Thompson's remarks, and the specific reference with the noting of the page number (1992: 333) seemingly concludes them. However, the words that follow are not the beginning of a new paragraph, and consequently it is difficult at first to distinguish whether this is a further idea, one that should be attributed to the same source, or whether this idea should also be ascribed to Thompson. A possible interpretation is that this is a formulation that combines both the view of Thompson and that of the author. If this is indeed so, the use of the first-person pronoun may indeed be an expression of the combination of the two views and a reflection of their consensus regarding this claim.

However, the right interpretation comes up only in later remarks. The next paragraph begins with an *adversative* (Fløttum et al. 2006) and confronts Thompson on the background of new findings:

(15)   *However*, on the basis of an analysis of the archaeological material now avail-
       able, *we* can revise and supplement this picture of Jerusalem.   (p. 151)

In the following sections, which discuss later periods, the author also mentions professional confrontations among archaeologists related to the dating and interpretation of various findings. She is not involved in all of these confrontations, and nor does she state her opinion on all of them, for example:

(16)   Unfortunately, no agreement has yet been reached on the dating of the pottery from this period (see, among others, Holladay 1990; Wightman 1990; Ussishkin 1996). (p. 151)

(17)   Jerusalem was then about 50 ha in size, although its precise size and limits are still being debated (Tushingham 1979, 1987; Geva 1979; Gibson 1987; Bahat 1993; Chen, Margalit and Pixner 1994) [...]. (p. 156)

(18)   Recently an attempt has been made to redate the inscription in the Siloah Tunnel to the Hasmonean period (Rogerson and Davies 1996; against this Hendel 1996; Hackett *et al.* 1997). (p. 159)

This point of dissent regarding the dating of archaeological findings lies at the heart of the dispute between Finkelstein and Mazar, which I will present in Section 5.5.

In contrast with Steiner's uninvolved stance in the later disputes, in other cases she injects herself into points of contention, sometimes very clearly presenting her own position:

(19)   This practically rules out the possibility that this cave was a favissa, a repository for objects used in a temple, as the excavator deduced (Kenyon 1974: 138; Stager 1982) (p. 155)

(20)   Shiloh (1986) interpreted this as a state archive, but the bullae were found among broken household pottery [...] thus an interpretation as a private archive *seems more plausible.* (p. 159)

In these two examples, the criticism is personal in the sense that the names of the researchers are explicitly noted. However, in Example (20), the hedge 'seems hardly possible' lends the criticism an indirect character.

The following example clearly demonstrates how the sense of confrontation is toned down: Steiner refers to herself in a non-integral reference in parentheses, thereby diminishing the status of her own position. The choice of this strategy creates only an implied confrontation:

(21)   Cahill and Tarler maintain that [...]. They date both structures in Iron Age I (1993, against this M. Steiner 1994). (p. 153)

Later, she draws the reader into her argumentative process by means of a question:

(22)   Who exercised this control is an interesting question. Was it the royal court supported by a large bureaucracy, or was it rather the urban elite of traders and artisans? (p. 160)

The Conclusions section of the paper is very short and it mentions again the difficulty in reconciling the archaeological evidence with the textual materials. The final sentence in the paper relates to these two types of sources:

(23)   Prompt publication of the excavated remains and new interpretations of relevant texts are urgently needed.   (p. 161)

The role of this sentence in the creation of dialogicity is especially interesting. Through it, the researcher appeals to the members of her research community, as she notes their shared endeavor and goals. She presents her own paper as part of this undertaking, which adds knowledge to the shared reservoir of knowledge but that leaves unanswered questions that other members of the community may be able to answer in the future. In doing so, she presents herself as a valuable member of the discourse community, whose contribution to the shared endeavor enables its advancement through dialogue with other members of that community.

In summing up the dialogic nature of this paper, it appears to hold a lively dialogue with the entire discourse community. It contains a large number of references; researchers who provided "hard" factual knowledge (archaeological findings, in this case) along with researchers who have drawn conclusions from this knowledge and scholars who suggested evidence from other branches of research are noted. Also mentioned are confrontations and disagreements in the research, and the author herself confronts some of the arguments she cites from others. Nevertheless, these confrontations do not lie at the center of the paper and the overall tone is restrained. She even ultimately makes a general appeal to the discourse community, one that calls for unity rather than discord.

Certain linguistic devices explicitly act to minimize confrontations:

The names of the researchers that she is opposed to generally appear in non-integral citations, in parentheses or even in a cluster of references, which reduces their salience and prevents a feeling of direct confrontation. Researchers that hold views opposed to hers are often noted without mentioning their names, as part of general expressions, such as *in most books* and *most of my colleagues.*

The presence of the author in the text is also restricted: She even cites herself in a non-integral citation in parentheses, thereby diminishing the status of her own position. She does not become involved or take a position in every disagreement mentioned, and even when her position is clear, it is often not explicit. The first-person singular pronoun appears only once.

The first-person plural pronouns in the article are inclusive and can be interpreted as referring to the entire community of researchers to which the author belongs. Consequently, they create more of a feeling of partnering than confrontation. Together with the use of the indefinite pronoun 'one,' and the passive voice, they create the rhetoric of objectivity that is typical of scientific writing. In the excerpt cited below, Biber et al. (1999) connect the use of all three of these elements as being interrelated:

> The relatively high frequency of the generic *one* in academic prose should be compared with the unexpectedly high frequency of *we* [...], and with the very high passive frequency in this register [...]. These are all connected with the preoccupation in academic work with making generalizations and with the wish to adopt an impersonal, objective style.
>
> (Biber et al. 1999: 355)

The dialogicity of the author with the reader in this paper is created by means of a number of linguistic means. The use of the inclusive 'we,' the indefinite 'one,' obligation modals and questions that on occasion are posed to the readers draw them in into the conceptual processes that underlie the argument.

In all these senses, this paper is an example of "classic" scientific dialogicity, which can be found in all disciplines. On the background of the dialogic nature of this paper, the following sections will demonstrate other kinds of dialogicity that are more direct, more personal and less reticent.

5.4 The conflicting pattern: Targeting a school (Na'aman 1996)

The paper that I will analyze in this section (Paper 2 in Appendix B) will be defined as a confrontational paper, although it is not part of what Hunston (2005) calls a 'conflict exchange.' It is not a response to a specific paper or researcher; its goal is rather to respond to the ideas of a group of scholars.

Nadav Na'aman is an Israeli historian from the department of Jewish history in Tel Aviv University. In this paper, he directly targets the status of Jerusalem in the tenth century BCE. As noted, the background for the disagreement is the absence of any archaeological findings in Jerusalem to testify to the strength or even settlement of the city during the relevant period. To support his argument, Na'aman enlists the Amarna letters (which I discussed in the analysis of Steiner's paper, above), and his declared goal is to offer a logical argument by way of comparison: The argument according to which there was no settlement during the period of the Amarna letters is based on the absence of archaeological findings from this period. Na'aman calls this argument *argumentum e silentio* – an argument from silence; however, the testimony of the Amarna letters contradicts this claim. According to Na'aman, the Amarna letters should be preferred as evidence of the existence of Jerusalem as a significant city during this period, with other means used to explain the absence of archaeological findings from Jerusalem. By analogy with this, says Na'aman, the same logic can be applied in regard to the period of the United Monarchy (tenth century BCE): The absence of archaeological findings contradicts the textual evidence, but the conclusions that certain scholars sought to draw from this fact are based on *argumentum e silentio*, and are consequently mistaken.

According to the way in which the author defined himself and his goal, his aim in this paper is not to make his own contribution to the existing scientific knowledge, but rather to explicitly express a position in an argument on a subject over which a conflict has been raging. From this it follows that the dialogue we can expect here is not merely a dialogue with the world of research, but rather a direct confrontation with opposing views along with an effort to refute them. We will see below how this is expressed by means of the various dialogic components of the discourse.

The dialogue with the discourse community is held in this paper by means of citations from the research literature so as to reflect its confrontational nature. Most of the references (80 percent) are of the non-integral type, including numerous clusters of references in parentheses, especially the claims made by opponents, which are thus presented in general terms. A number of citations of texts in quotation marks appear, in most cases to present claims that the author is opposed to. Thus, for example, about half of the Introduction section is devoted to the citation of arguments made by researchers that he calls "revisionists" (these are the "minimalists") that the author plans to refute, with similar claims presented at length later on as well.

As noted, the introductory section contains a long list of citations, some in general terms (non-integral) with others as integral citations that contain large chunks of text. All the claims cited in the introduction are those to which the author of the paper will present opposition later on. This design of the paper marks out its confrontational nature from the outset. This nature is also evident from the way in which the presence of the author is underscored by means of the use of the first-person singular pronoun:

(1) Let *me* cite a few examples of the conclusions reached by some advocates of this "revisionist" school of thought.   (p. 17)

(2) *I* will limit my criticism to two major points [...].   (p. 18)

This is what Martín-Martín calls writer-mediated criticism. The final sentence of the Introduction also begins with the first-person singular pronoun and contains a strong emphasis on the paper's confrontational intent.

(3) *I* will try to show that these scholars' evaluation of the excavations in Jerusalem is *inadequate* and leads to *erroneous* conclusions [...].   (p. 18)

The adjectives 'inadequate' and 'erroneous' already contain direct criticism of the cited researchers' claims. The same is true for the adjective 'illegitimate' in the heading of the next section: "Excavations of the Ophel Hill: Legitimate and illegitimate conclusions." The reference to illegitimate conclusions in the title is already indicative of the confrontational nature of the section. While it begins with citations from archaeologists in order to present findings as background for the discussion, later claims of researchers are cited in order to present opposition to them. These scholars are those who, in his view, drew mistaken conclusions based on the fact that no archaeological findings were found from a particular period, i.e. based on *argumentum e silentio*. The rejection of this argument is carried out by means of an appeal to shared knowledge (Hyland 2001, 2005) presented as shared by the research community; consequently, the fact that those researchers ignored it cannot be justified.

(4)   *As is well known*, conquest, destruction and desolation leave distinct marks that archaeologists can easily expose; uninterrupted continuity of settlement, on the other hand, leaves only a few remains of the earlier building activity.
(p. 19)

In this way, the author holds a dialogue with the discourse community and distances those researchers that in his opinion disregard this shared knowledge from the community.

The question of the legitimacy of the opponents' conclusions is also raised by means of an indirect question:

(5)   In light of this discussion, *one* may ask if it is legitimate to draw negative conclusions about tenth century Jerusalem on the basis of the archaeological excavations conducted at the site.   (p. 19)

The formulation of the question and the use of the indefinite 'one' are of dialogic significance: In this way, the author leads the reader to ask this question and aspire to find the answer. Although in the absence of the first-person pronoun, the criticism is non-mediated, the author's negative answer (that it is not legitimate) is provided immediately afterwards in the subsequent text.

This is where the main thrust of the argument is first presented. At first, the author relates to the period of the el-Amarna letters and the contradiction between what is described in them and the archaeological findings from Jerusalem during this period. The scholars cited in this part are those who drew conclusions from this contradiction, conclusions that in his view are mistaken and implausible. Among them, he mentions Steiner (see above):

(6) The paucity of LB II remains led Franken and Steiner (1992) to doubt the identification of the Amarna city of Urusalim with Jerusalem. [...] The two scholars were unaware of the problems of excavating this highland site, and drew illegitimate conclusions on the basis of negative evidence.   (p. 19)

Here too, then, the citations take on a confrontational role – the cited scholars are those to whose views the author is opposed. The author's criticism here is personal and direct.

The place of the researcher in the design of the logical structure of the argument is underscored by means of the use of the first-person singular pronoun:

(7) To clarify this problem I will first present two archaeological instances from other periods in the history of Jerusalem, and then will draw the *logical* conclusions.   (p. 19)

The adjective 'logical' belongs to the set of positive attributes, and he uses it to characterize his own conclusions, as opposed to those of his opponents, which he previously characterized as 'inadequate', 'erroneous' and 'illegitimate'.

The next section of the paper is devoted to a detailed presentation of the situation as it emerges from the Amarna letters with a resolution of some of the questions that arise from it. In contrast to the previous parts of the paper, the citations in this section serve to support and describe the reality as reflected by the letters.

The presence of the author is underscored when he describes the structure of his argument using the first-person singular pronoun:

(8) In what follows *I* will suggest that comparing of the political formations [...].
      (p. 19)

However, this section also contains a number of first-person plural pronouns as well, such as:

(9) The discussion will focus on the central highlands of Canaan in LB II, since only for this period do *we* have both contemporaneous documentary evidence [...] and detailed archaeological data [...].   (p. 19)

The author creates a dialogue with the research community through the use of these first-person plural pronouns, thereby relating to the shared knowledge and shared goals of the community. Especially interesting is the use of the first-person pronoun as it relates to the action of drawing conclusions:

(10) *We* may conclude that the Canaanite rulers were considered alike in intrastate relations, in internal relations within Canaan, and with relations with their subjects [...].   (p. 21)

In this way, the author shares his thought processes with the readers, leading them through the various stages of his argument so that they are led to draw the same conclusion as the author.

In two places, other researchers are referred to by means of the general designation 'scholars':

(11)   On the basis of the archaeological evidence alone, *scholars* would have assumed that a few small rural communities [...] lived in the highlands of Judah and Benjamin in LB II. However, in light of the documentary and archaeological evidence, the city of Jerusalem may be defined as a highlands stronghold [...].   (p. 20)

(12)   On the basis of the archaeological evidence alone, *scholars* would have suggested that LB II Shechem was the center of a medium-sized Canaanite city-state [...]. Only the evidence of the Amarna tablets enables us to correctly evaluate the historical reality of the time.   (p. 20)

These two examples contain concession structures that the author uses to contrast the assumed conclusions of those unspecified scholars with his own.

Here too, the author mentions knowledge shared by the community in such a way as to provoke criticism of the opposing researchers and present the author and reader together in a preferred position as those that are aware of the shared knowledge and who draw the proper conclusions accordingly:

(13)   With these conclusions in mind, it is important to note the difference between the self-perception of the people of the Late Bronze Age and modern scholarly definitions. The importance of studying an ancient civilization in its own terms and within its own system of values *has been commonly accepted by scholars* since Landsberg's seminal study (1926) on the conceptual autonomy of the Babylonian world.   (p. 21)

The appeal to the reader by means of an obligation modal is demonstrated in Example (14):

(14)   This discrepancy between contemporaneous concepts and our modern definitions *should* be taken into account [...].   (p. 21)

The next section in the paper is introduced with a presentation of Jamieson-Drake's sociological model, which the author accepts. His support for this view is accentuated by the use of a number of adjectives that represent positive disciplinary values: 'clear', 'reasonable', 'sound':

(15)   He suggests *clear* definitions for a chiefdom and for a state [...] His conclu-
        sions are *sound* and *reasonable* and accord well with the survey conducted
        recently in the area of the kingdom of Judah (Ofer 1994).   (p. 21)

The coherence with others' findings is also a particularly important point, which
strengthens the validity of the conclusions (see Section 3.2 above).

Immediately afterwards, the author distinguishes Jamieson-Drake's position
in principle from those whom he calls "his followers," those researchers who drew
"farfetched inferences" from what he wrote. Those researchers are revisionists,
whose position is presented in a generalized recapitulation, and whose names appear in a non-integral reference, in parentheses and in a cluster of references. The
author's long response to them is supported by a number of citations that bolster
his own position, including self-citations. The positive adjective 'clear' appears
here in a contrastive structure, which underscores the author's reservations with
their claims:

(16)   While Jamieson-Drake draws a *clear* line between biblical and archaeological
        definitions and data, the line is *less clear* in the works of his followers. (p. 22)

The author's dialogue with the research community is constructed in an interesting concession structure:

(17)   *I* agree with Van Seters (1983: 309–310) that the description of the temple
        with all its appurtenances is the work of the Deuteronomistic historian and
        reflects the temple of the late Judaean monarchy. *Yet* the memory of Solomon
        as builder of the temple in its initial stage must be authentic, and it is even
        possible that the historian had seen a Solomonic building inscription from
        the dedication of the original temple.   (p. 22–23)

In the first sentence, the use of the first-person singular pronoun and the mention
of the name of a specific researcher with a precise reference present the discourse
as a dialogue seemingly held between two scholars, although it is entirely possible
that the scholar cited here is a representative sample of other scholars who hold a
similar position. Nevertheless, it is the explicit dialogicity with the other researchers that underscores the author's claim as one that is presented from a position of
personal authority and of valuable membership in the discourse community.

The first-person plural pronouns, on the other hand, create a sense of solidarity with the discourse community by mentioning its shared knowledge and the
ability of the members of the community to draw conclusions from it:

(18)   *We* may further assume that, like all Syro-Palestinian rulers, those of the
        Jerusalemite dynasty were considered kings both by their neighbors and by
        the inhabitants of the kingdom.   (p. 23)

The following example, however demonstrates the power of the first-person plural pronoun in leading the reader through the different steps of the argument:

(19)   This brings *us* to the problem of the sources of the study [...].   (p. 22)

Na'aman's paper continues by describing in the next section the findings not from Jerusalem, but rather from the areas around it. The section begins with a question, the function of which is to introduce the new subject:

(20)   What was the number of settlements and the scope of population governed from the new court of Jerusalem? Recent surveys supply this essential information [...].   (p. 23)

Another question is presented in this section at the beginning of a paragraph, in order to introduce a new subject that the author immediately develops as he continues:

(21)   What happened in the stronghold of Jerusalem after the rebellion and the establishment of the northern Kingdom? According to the (yet unpublished) new surveys [...].   (p. 24)

However, at the heart of the section lie two rhetorical questions, the first of which is:

(22)   Some "revisionist" scholars assume that it is impossible that such an "empire" (i.e. great kingdom) was conquered by David and governed from Jerusalem [6 references]. But is it really impossible? [6 sentences omitted]. There is, therefore, nothing impossible about the main outline of the biblical account of David's conquest.   (p. 23–24)

The citations from revisionist scholars are presented here in order to refute them. They are presented in general terms and the names of the researchers are once again presented as part of a cluster in parenthesis, which demonstrates that this is not a confrontation with a particular individual scholar, but with a group. Distinguishing its various members is less important to the author than formulating their shared claims. The wording of the question ("is it really...?") already implies that the author does not believe that it is impossible, i.e. the question does not appear to be a genuine one, but rather is a rhetorical question, the answer to which is negative. And indeed, as expected, immediately after the question, he devotes six sentences to raising counterclaims, following which he presents the negations of these claims as the conclusion of his discussion. The rhetorical question is a persuasive device directed at the audience, to stimulate and enable it to participate in the implied dialogue with the speaker.

The response of the author is supported by a long citation from a scholar named Liberani in a separate paragraph, which is followed by an utterance that further supports it:

(23)   This ideological concept, which is supported by many historical examples, may well explain the biblical description of David's great kingdom.   (p. 24)

The author then emphasizes the coherence of the cited approach with other, well-known phenomena, and attributes to it explanatory power, which also provides support for his position. Later in the section, three long consecutive citations are presented from Thompson (1992), one of the chief targets of his attacks, followed by yet another rhetorical question formulated as an indirect question:

(24)   *We* may ask on what Thompson bases these far-reaching conclusions. He does *not* discuss the archaeological data in detail, *nor* does he examine the *longue durée* or compare the Iron Age IIA data to other documented periods. His conclusions *neither* take into account all the available evidence *nor* do justice to the complexity of the problem.   (p. 24)

The question does not receive a direct answer in the text; however, from the wording of the sentences that follow, it is clear that the author believes that Thompson's conclusions are unfounded. This is then one of the most common forms of the rhetorical question: The claim that it implies is diametrically opposed to the formulation of the question: "On what does he base himself?" = "He does not base himself on anything." The rhetorical question and the discussion that follows it positions the opponent as one whose words have no real scientific foundation.

It is interesting to note here the considerable use of sentences containing negation, which is a well-known polemic element (Fløttum et al. 2006: 244–245). The rhetorical question as an element that creates dialogicity is joined here by the use of first-person plural pronouns, which creates a sense of the reader sharing in the author's thought process. The first-person singular pronoun, on the other hand, which appears immediately afterwards, underscores the presences of the author in the text and argument:

(25)   On the basis of the data presented above, *I* would suggest that Judah in the late tenth–ninth centuries B.C.E. was a peripheral small and powerless kingdom […].   (p. 24)

In this way, after the dismissal of Thompson's position as groundless, the author's view is given preference based on an authoritative stance.

The first-person plural pronoun appears a few more times in this section. In all, it relates to cognitive actions that the author carries out as he shares them with the reader too:

(26)   *We* may assume that following the division of the monarchy, Jerusalem governed about 35–40 highland sites [...].   (p. 24)

(27)   Even if *we* assume for a moment that the United Monarchy encompassed only the highlands on both sides of the Jordan [...].   (p. 23)

(28)   By way of comparison, *we* may recall the offensive of Lab'ayu, king of Schechem.   (p. 23)

The final section in the paper contains a series of conclusions, the last of which relates to the question of the definition of a *state*. This conclusion is formulated by means of the concession structure:

(29)   Scholars must always take into account the gap between *our* modern definitions of states and societies and the self-perception of ancient societies. [...] Modern definitions may be the more "scientifically" accurate, but it is equally important to analyze ancient societies according to their own terms and self-perception.   (p. 25)

The author turns here to the entire research community, as arises from the use of the term 'scholars' and the absence of any mention of specific researchers. His dialogue is held with the members of the community and with its accepted conventions and norms. The claim that appears in the satellite of the concession, that "Modern definitions may be the more 'scientifically' accurate," hints at the need of researchers in the humanities to appear scientific by aspiring to scientific precision in those areas in which this kind of precision is difficult to achieve. By means of the quotation marks around the word "scientifically," the writer expresses his reservations with the possibility of arriving at the truth by means of criteria that in his view are inappropriate, which may be borrowed from other areas of knowledge. However, he attributes the scientific definitions not only to other members of the community, but also to himself, with the proof being the use of the first-person pronoun 'our' in the first sentence. The referent of the pronoun 'our' appears to be very general – involved are definitions that the world of science as a whole subscribe to. Through the use of this formulation, the author does not exclude himself from the community; on the contrary, he emphasizes his membership in it. It is this speaking from within the community that enables him to suggest that others (and perhaps himself too) should be wary of taking rash and dangerous steps that could distance the entire scientific community from the truth. This plural form can serve, according to Myers (1989) as a politeness device. Myers maintains that one way of making a criticism in a polite manner is for the writers to use pronouns that include themselves in the criticism (Myers 1989:7).

The frequent use the author makes of quotation marks for the purpose of expressing reservations in the paper is noteworthy. Certain words appear regularly in quotation marks, for example the adjective "revisionist" in Examples (1) and (22) above. At its first appearance in the paper, one might reasonably assume that this is just his way of marking a new term, but later in the text, it becomes clear that the quotation marks appear every time this school of thought is mentioned. In fact, the adjective "revisionist" is one that the members of that school would use to describe themselves. Thus, the author's use of the quotation marks seems to signify his reservation with the use of this word.

Additional words that appear in this paper in quotation marks are "modern," "objective," "scientific," and this is somewhat surprising. Let's look at the relevant contexts:

(30)   By analyzing the results of the archaeological excavations and surveys conducted in the areas of the kingdom of Judah, and using "modern" sociological definitions, they suggest that Jerusalem became the center of a state no earlier than the eighth century B.C.E.   (p. 17)

Later, he explains his reservations with these "modern" definitions:

(31)   In contemporaneous concepts, the territorial highland entities were regarded as kingdoms and their rulers as dynastic kings. Anthropologists and sociologists, on the other hand, use more "objective" criteria for defining such entities.   (p. 21)

The objection to the adjective "objective" is surprising since objectivity in science is a positive value that is very difficult to mock. However, placing the adjective "objective" in quotation marks shows that in the view of the author, it is inappropriate to define the relevant criteria as objective. This also explains his reservations with the "modern" definitions (Example (30)), which are "objective." The word "modern" appears in quotation marks twice in the paper.

To this can be added the appearance of the adverb "scientifically" in quotation marks, once again in the context of sociological definitions, as in Example (29) above. In view of the fact that the terms "scientific," "objective" and "modern" possess a clearly positive value in scientific contexts, reservation with them can be understood only on the background of a conflict between the positions. The "scientificness," "objectiveness" and "modernness" all refer to the opponents whose views are rejected in this paper. Through the use of the quotation marks and reservation with them that they indicate, the author presents them as values that are being used in a distorted and improper manner. This then is yet another linguistic device to create confrontational dialogicity.

The dialogicity in this paper is consequently of a different nature than the "classic" dialogicity as shown in Steiner's paper in the previous section. The paper's entire and explicit aim is to attack and refute specific positions and claims and express a different one. The attacks on those holding opposed views are of different degrees of directness, and numerous linguistic elements are employed to this end: the formats of citation, the choice of pronouns, concession structures, questions and particularly rhetorical questions, quotation marks to express reservations and explicit utterances of rejection, such as the quite unambiguous adjectives in the following phrases: "inadequate evaluation," "erroneous conclusions," "illegitimate conclusions" and "farfetched inferences" which undermine the ethos of the opposing scientists.

All of these elements combine to create a text whose confrontational nature is direct and personal and whose dialogicity is far more explicit than that expressed in the "classic" scientific dialogicity.

5.5 The conflicting pattern: Targeting a researcher (Finkelstein et al. 2007)

The paper that I will analyze in this section is a response to a paper by Eilat Mazar (2006, Paper 6 in Appendix B). In order to understand the response, I will first present Eilat Mazar and her relevant papers on the subject.

Eilat Mazar is an Israeli archaeologist from the Hebrew University of Jerusalem. In the years 1996–1997, she published a hypothesis in two papers (Papers 3 and 4 in Appendix B) according to which the palace of King David might be found at a certain spot in Jerusalem, which had yet to be excavated up until that time. She based her hypothesis among other things on findings from previous archaeological excavations, the topography of the area and the characteristics of settlement development of the city. However, the main basis for her hypothesis appears to be the biblical text. Of special interest is the paper in Hebrew (Paper 3 in Appendix B), which is in fact a lecture that was given at a conference held in Israel in 1996 and published in the conference proceedings. The paper begins with a presentation of the details of the biblical story in such a way as to convey them as historical details:

(1)   Hiram King of Tyre built a palace in Jerusalem for King David to symbolize the covenant of friendship between them and as a sign of his recognition of the new status of Jerusalem as the capital of the United Monarchy.

                                                                 (E. Mazar 1996:9)

Her comments later in the paper also demonstrate that she does not challenge the biblical story or any of its details, for example:

(2)   It should be borne in mind that when David conquered the city, he left its Jebusite inhabitants alive.   (E. Mazar 1996: 12)

Many additional details, which serve the author as the basis for her hypotheses, are reported in this way in the paper. The most important point in the biblical text for Mazar's hypothesis is the verb "went down" (Hebrew *yarad*) that appears in II Samuel 5:17: "David heard of it, and went down to the stronghold." This is how Mazar reports her conclusions from this verse:

(3)   When David heard that the Philistines were arising to fight against him, he went down to the stronghold. From this, we can assume that the place where David had been when he heard about the Philistines' camping near the city was higher than the stronghold.
      There is no reason to doubt the precision of the biblical description of the situation as it was, similar to the biblical description of the opposite situation, when David hears of the death of Absalom: "And the king was much moved, and went up to the chamber over the gate, and wept" (II Samuel 19:1).
      It may be assumed then that David was in Jerusalem in his home – his new palace whose construction had just been completed, and when he heard about the arrival of the Philistines, he went down to the stronghold. From this it follows that David's palace was in a place that was higher than the stronghold to which he descended. Moreover, it is reasonable to assume that the palace was not sufficiently fortified to withstand a direct attack, and that consequently, the king preferred to entrench himself in the stronghold until the danger had passed.
      If the palace was indeed situated higher than the stronghold, the question remains as to where the palace must have stood.   (E. Mazar 1996: 11)

It is clear that Eilat Mazar's hypothesis regarding the location of the palace is based on a literal reading of the biblical text and her acceptance of it as reflecting historical truth in every detail.

   Mazar's paper in English from 1997 (Paper 4 in Appendix B) is not a peer-reviewed scientific paper in every sense. It was published in the *Biblical Archaeology Review* (BAR), a semi-popular journal. Many of the articles published in this journal are commissioned by the editor. He adds headings of his own and interjects his own comments for the sake of the readers, including details about the authors, the context and various subjects that arise from the article, such as photographs and illustrations. Her paper in this journal Mazar concludes with an appeal, which is also a request for support to help finance the excavations:

(4)   Who will heed the call to find King David's palace?   (E. Mazar 1997:74)

After she managed to raise the necessary funding, Mazar began to excavate in 2005, and in 2006, she published her first findings. The excavations uncovered the remains of a large stone structure, which she dated to 1000 BCE, and which she claimed was likely to have been the palace of King David. She presented her findings to Israeli archaeologists at another conference of the *New Studies on Jerusalem* series, and published them once again in the proceedings. In this paper (Paper 5 in Appendix B), she gives a detailed description of the findings uncovered up to that point. Following the description of the large stone structure, Mazar offers certain hypotheses regarding its identification and then proceeds to reject them. The paper concludes with this paragraph:

(5)   The explanation that remains, the one that is most plausible, is that this is King David's new palace. The historical biblical description (II Samuel, Chapter 5) of David and his Phoenician allies, the highly reputed builders who build his new palace infinitely correspond with the facts uncovered in the excavation of the large stone structure briefly described here.   (E. Mazar 2006a: 16)

In the years 2006–2007, Eilat Mazar published three more papers along with an archaeological report related to these findings. The paper that was published in the *Biblical Archaeology Review* (Paper 6 in Appendix B) is of a more personal and sentimental nature, and it mentions her late husband, her children and in particular her grandfather, Benjamin Mazar, who was the president of the Hebrew University and an eminent archaeologist who excavated in Jerusalem. When she was just starting out, she worked with him and published findings from his excavations. In this article, she reports on her conversations with him, and thus, for example, she writes the following about her grandfather and how he influenced her:

(6)   One of the many things I learned from my grandfather was how to relate to the Biblical text: Pore over it again and again, for it contains within it descriptions of genuine historical reality.   (E. Mazar 2006b: 20)

Mazar's claims regarding the finding of David's palace naturally received attention that went beyond the boundaries of academic publication and were published in Israeli newspapers. In the first interview she gave to the press, four days after completing the first season of digging, her findings were described as "a discovery that is poised to shake the world of archaeology." Mazar herself is quoted in the interview as saying: "This is not just another discovery. It is on the scale of a miracle."

At the end of the second season of Mazar's excavations, four archaeologists from Tel Aviv University published a paper responding to her claims (Paper 7 in Appendix B). It is this paper that I will analyze here as yet another example of a paper in the confrontational pattern. The examples in this section from this point forward come from this paper, which was published in the journal of the Department of Archaeology of Tel Aviv University, which is a peer-reviewed journal. However, it is notable that two of the authors, Israel Finkelstein and David Ussishkin, are on the editorial board of the journal and consequently, it may be assumed that the paper did not face considerable challenges to acceptance for publication.

The confrontational nature of the paper is evident already in its title, which is worded in the form of a question: "Has King David's palace in Jerusalem been found?" This title in fact echoes the title of Eilat Mazar's paper from 2006 (Paper 6 in Appendix B): "Did I find King David's palace?" As noted, Mazar's answer to this question was apparently yes. The echoing of the question in the title of the response paper hints that the authors intend to examine the very same findings and offer a different answer. They describe their motivation for writing their paper towards the end of the Introduction. The use of the critical expression 'media frenzy' is notable:

(7) The ostensible importance of this discovery and the *media frenzy* that has accompanied the excavation demand immediate discussion [...].   (p. 142)

The first section of the paper presents a brief historical survey of the archaeological research at the relevant site, with quotes from a number of archaeologists. These quotes will serve later to refute Mazar's claims.

The following section is devoted to Mazar's findings. The heading of the section "The findings according to Eilat Mazar" hints at the author's reservations with these findings, or at least with the interpretation Mazar offered to explain them. In this section, they detail the findings in a somewhat different fashion from that in which Mazar presented them, and describe how Mazar proposed that they be interpreted.

In this context, there are two citations from Mazar – one is short and places only two words in quotation marks. It is stated somewhat casually, but will take on importance later, at the stage when the claims are refuted:

(8) According to her the main wall of this building (wall 107), described as "slightly curved," runs from west to east [...].   (p. 144)

Later, the authors relate to this point and explain that the wall seems "slightly curved" because it "in fact represents two separate structures: [...] The western part of the wall is built in a straight line [...]. The eastern part [...] forming a winding line that runs diagonal to the western part of the wall" (p. 155). They thus suggest a completely different interpretation from the one proposed by Mazar, and accordingly, the quotation marks surrounding the phrase "slightly curved" may even be a marker of irony or mockery.

A mocking tone can be heard in yet another citation from Mazar in this section:

(9)   According to her observation, an "arched cistern is located at the western end of our excavation area, its arch having been built into W107 of the Large Stone Structure… The impression already received is that the cistern was hewn in the earliest stages of human activity in this area. Once the Large Stone Structure was built (… The Iron Age IIA) *the cistern was incorporated into the structure*, and would subsequently be used continuously in the preceding [*sic!*] periods of activity" (E. Mazar 2007:73, our emphasis).   (p. 145)

The long quote, the emphasis of part of it that apparently hints at the implausibility of the claim in the view of the authors and the pointing to a mistake in the text all together create a mocking tone in this quote too.

The following section of the paper is headed "The finds: An alternative interpretation," and it is dedicated to refuting Mazar's claims. Already at this stage, the authors present their conclusion, which relates to Mazar's insistence on a literal reading of the biblical text:

(10)   Mazar's theory seems to have been affected by her own interpretation of the biblical verses rather than based on factual data (see discussion below).
                                                                                      (p. 148)

The wording clearly undermines Mazar's scientific ethos, suggesting that she ignores hard "scientific" facts. The remark in parenthesis refers the reader to a later part of the paper, and this point will in fact be the subject of development in a separate section towards the end of the paper.

This section, which offers "an alternative interpretation of the findings" is the essence of the paper. It is filled with utterances expressing reservations with various details of Mazar's interpretation, such as:

(11)   Therefore, although the dating of the cupmarks to the Chalcolitic period is possible, it is not the only possibility.   (p. 147)

(12)   As for Argument 2, Mazar's interpretation, according to which the 'Stepped Stone Structure' and 'Large Stone Structure' belong to one Iron IIA complex, is based on circumstantial considerations that are open to alternative interpretations.   (p. 149)

(13)   The basket numbers from this place apparently disclose that at least one Iron IIB item […] was found under four (of the seven) Iron IIA items (from basket 661)! The division between lower Iron IIA and higher Iron IIB material

within Locus 47 is therefore questionable, and so is the idea that this is an *in situ* assemblage. (pp. 149–150)

The exclamation point, something we encounter only very infrequently in the scientific discourse, denotes that the authors view the fact that has been described as irrefutable proof of the weakness of Mazar's claims. It may also be a sign of the amazement the authors feel at the fact that Mazar herself did not draw the right conclusion from this fact. In this context, a reservation raised by Mazar herself is cited at length, and this reservation is employed here for the purpose of refutation:

(14) Significantly, the excavator herself doubts whether the pottery in Locus 47 was found *in situ*, and suggests that it is part of a fill that had been brought here from elsewhere: "The state of preservation of the vessels suggests that they were at some nearby location prior to the construction of W22 and W24, and somehow were deposited at this spot when the walls were built" (*ibid.*: 61).

(p. 149)

In another place, we can see how Mazar's own words are used against her:

(15) Several of Mazar's finds indicate the possibility that walls of the 'Large Stone Structure' were built in post-Iron Age times: A late Iron II bulla was found "tucked away among the masonry of the Large Stone Structure" (E. Mazar 2007: 19). (p. 150)

Other quotes in this section are from other researchers, which are cited in order to support the claims of the authors. Let us look for example at the structure of the following paragraph:

(16) *The common view* assumes that the 'Stepped Stone Mantle' [...] had been erected in a single construction effort [...]. A *fresh* examination by *us* in the field and a reassessment of the published data lead *us* to conclude that there are significant structural differences between the lower and upper parts of the 'Stepped Stone Mantle', and that it represents more than one phase of building activity. The recurrent need for a revetment in this spot was first observed by Duncan, who discovered a fissure in the rock scarp: "The hollow or fissure referred to is filled up with great boulders [...]" (in Macalister & Duncan 1926: 52). (p. 151)

The paragraph begins with a non-specific reference to other researchers ('the common view'). This is followed by the authors' reservations with that same common view. This reservation is underscored by means of the positive description 'fresh assessment' and the authors' twofold use of the first-person pronoun 'us' to refer to themselves, which runs counter to the non-specific nature of the opposing view. The authors' reservations are further supported by a specific quote from one of the first archeologists to excavate in Jerusalem.

Another example of citations that support their view can be seen in the following paragraph:

> (17)   Regarding the upper part of the 'Stepped Stone Mantle', […] it must have been constructed – or at least rebuilt – in the late Hellenistic period. This date was originally suggested by Kenyon (1974: 192–194). Shiloh also observed the association of the upper part of the 'Stepped Stone Mantle' with the Hellenistic fortification system: "the line of the 'First Wall' and its towers integrated the top of the stepped stone structure…" (Shiloh 1984: 30).   (p. 154)

The refutation is supported in this section by a number of concession structures. Let us look at a paragraph based on two concession structures, which present the authors' reservations in view of the data or interpretations attributed to other scholars:

> (18)   Two pieces of information possibly indicate the date of the lower part of the 'Stepped Stone Mantle'. First, the latest sherds […] date to the Iron IIA (Steiner 1994: 19; 2001: 50; 2003: 358). *However, as the connection of these structures* to the 'Stepped Stone Mantle' is not firmly established, this datum *should* be used with reservation. Second, Iron II houses were built over the lower part of the 'Stepped Stone Mantle'. Cahill argued that pottery […] was found on the two lower levels […] (Cahill 2003: 56–66 […]). *Yet*, only sherds, rather than complete vessels, were uncovered here, and they are registered as originating from ten different loci.   (pp. 151–152)

The concession structure creates the confrontational dialogue with the other researchers, and their claims are rejected as being insufficiently cautious. The obligation modal 'should' helps the authors to involve the reader in the reservation with the proposed interpretations.

In the Discussion section, the authors use yet another concession structure, in which Mazar's claims are explicitly rejected. Mazar's presence here is underscored by means of the use of her full name in the body of the text, in the position of the satellite of the concession, further emphasizing the reservation with her claims:

> (19)   *One* can argue – with Eilat Mazar – that the original 'Stepped Stone Structure' […] was constructed in order to support the slope and prepare for the construction of the 'Large Stone Structure'. The Large building blocks and the proto-Ionic capital […] could have collapsed from the building on the crest of the ridge (E. Mazar 2007: 54). *Yet*, dating all the walls of the 'Large Stone Structure' to the Iron IIA raises serious difficulties […].   (p. 154)

Here too, the indefinite pronoun 'one' helps to appeal to the assumed reader, who may not yet be convinced and who persists in accepting Mazar's interpretations.

Further in the paragraph too, the authors sum up the issue of the dating and further undermine Mazar's position by means of the concessionary structure:

> (20) Each of these problems can be explained away individually [...]; as a *set* of difficulties, *however*, they cannot be easily dismissed.   (p. 155)

The reader is thus led via the author's complex of arguments by means of the pronouns that ultimately lead him through the logical process:

> (21) This leaves *us* with a third alternative, according to which the walls of the 'large stone structure' were built in the Hellenistic period. But first *one* needs to examine whether the elements uncovered by Mazar indeed belong to a single structure.   (p. 155)

The indirect wording raises a question that the authors answer in the next paragraph in such a way as to completely dismiss Mazar's interpretation:

> (22) The plan of the 'Large Stone Structure' consists of three distinguishable elements (Fig. 1): Two described by Mazar as one feature [...]. As far as *we* can judge, wall 107 *in fact* represents two separate structures [...].   (p. 155)

The pronoun 'we' here is exclusive, refers to the authors, underscoring their opposing view and authority.

In the next section, which is entitled *"Eilat Mazar's 'historical' interpretation,"* the authors again bring up, this time in detail, their claim that Mazar insists on believing the biblical text rather than the archaeological evidence. Their reservation with her historical interpretation is expressed in the quotation marks that appear in the title. They define the principles of her method as follows:

> (23) Mazar follows two principles: (1) Biblical data are accepted without criticism as the basis for archaeological interpretation; (2) Therefore, biblical information takes precedence over archaeological data.   (p. 160)

They formulate their claims in such a way as to present her as one whose conclusions are incompatible with her very own findings, since they give primacy to the Biblical text.

> (24) As she admits, the chronological data recovered in her excavation indicate that the sole Iron Age fortification system extending in this area was in use during the 8th–7th centuries BCE. However, according to the biblical sources the Solomonic city-wall must have passed here, hence the fortification system in question must be Solomonic in date.   (p. 160)

They also accuse her of ignoring 30 years of research on the Book of Genesis and the patriarchal narratives, and disregarding decades of research on the Conquest tradition.

These harsh comments gravely undermine Mazar's ethos as a responsible and up-to-date scientist. Long citations from her paper are presented here to demonstrate her mistaken approach as they see it.

The brief Summary section contains the presence of the entire discourse community, its goals and aspirations to advance in the shared scientific endeavor. First, a sophisticated employment of the concessionary structure is used:

> (25) Eilat Mazar's excavations in the City of David add several points of information to what *we* know about the history of this problematic site. *Yet*, the main find – the 'Large Stone Structure' – was not properly interpreted and dated.
>
> (p. 161)

The satellite of the concession mentions Mazar's contribution to the shared scientific project, with the use of the first-person pronoun 'we' being inclusive and referring to the world of research as a whole and to all those who seek the truth. However, in the nucleus of the concession, the authors imply a rejection of the importance of this contribution. The meaning of the claim that the finding "was not properly interpreted and dated" is that the structure that Mazar uncovered is far less important than was previously thought.

The authors include the reader in their reservation through the use of the indefinite pronoun 'one':

> (26) Beyond archaeology, *one* wonders about the interpretation of the finds.
>
> (p. 162)

In the paper's final sentence, the authors relate to the existing trends in archaeological research, and once again state that Mazar's approach is archaic and obsolete.

> (27) This is an excellent example of the weakness of the traditional, highly literal, biblical archaeology – a discipline that dominated research until the 1960s, that was weakened and almost disappeared from the scene in the later years of the 20th century, and that reemerged with all its attributes in the City of David in 2005. (p. 162)

An analysis of all the various uses of first-person pronouns used in this paper demonstrates a number of interesting points. It is used to refer to the research community only once, in Example (25) above taken from the Conclusions chapter. Most of the uses of *we* refer to the authors themselves and relate to their own claims and conclusions, for example:

(28) In what follows *we* summarize her finds from bedrock to the Byzantine period.   (p. 143)

(29) *We* should note that there is no clue in the archaeological record for a Late Bronze reuse of the Middle Bronze fortification.   (p. 160)

(30) *We* find both suggestions difficult to accept.   (p. 147)

This is a writer-mediated criticism, in Martín-Martín (2005)'s terms, which strengthens the confrontational nature of the text. At the same time, some of the appearances of the first-person plural pronouns include the readers in the way the arguments are framed, and help them to follow the arguments in order to arrive at the same conclusions, for example:

(31) Before continuing, *we* must take a closer, fresh look at the 'Stepped Stone Structure'.   (p. 150)

(32) Turning back to the area of the 'Large Stone Structure' *we* see that Mazar reads the biblical references to Jerusalem in a sequential, literal way.

(p. 160)

In contrast, of particular interest are the first-person pronoun that portray the authors as those who saw the findings with their own eyes:

(33) [...] discussion, which is based on the preliminary publications and on *our* own observations made during *our* visits to the site in both excavation seasons.   (p. 142)

(34) This was also *our* impression when Mazar showed *us* selected sherds from this layer during a visit to the site.   (p. 148)

The intense presence of the authors in their text is interesting on the backdrop of the impersonal tone that normally typifies scientific writing. Impersonal writing serves the ethos of the scientist in that it reflects his objectivity and the separation between himself and the object of his research. However, whereas in the natural sciences, objectivity is a central value, and consequently the dominant tone is impersonal, in certain areas of the social sciences and the humanities, for example anthropology, the highlighting of the presence of the researcher as one who "was there" is what lends him his credibility and authority. The anthropologist Clifford Geertz (1988) discusses at length the place of the anthropologist author in his own texts and underscores the importance of the presence of the personal perspective in ethnographic descriptions:

> The ability of anthropologists to get us to take what they say seriously has less to do with either a factual look or an air of conceptual elegance than it has within their capacity to convince us that what they say is a result of their having actually

> penetrated (or, if you prefer, been penetrated by) another form of life, of having, one way or another, truly 'been there'. (Geertz 1988: 4–5)

According to the papers before us (this one and others that I will analyze), the field of archaeology also appears to accord importance to the physical proximity of the researcher to the findings, to his actual physical presence at the site of the excavations and to the fact that he saw the findings with his own eyes. His authority as a reliable scholar is therefore created by means of a balance between two forces: on the one hand, the fact that he has an "objective" approach, one that is free of emotion or interests, and on the other, his physical proximity to the findings. From a linguistic aspect, these are contradictory forces: The appearance of objectivity is characterized by minimizing the presence of the author in the text, whereas his authority as someone who "was there" requires an emphasis on that presence. These two forces and their importance to the discourse community will come up repeatedly in the coming sections, in the analysis of other papers.

To sum up, this paper too, like Na'aman's paper analyzed in Section 5.4, is of a deeply confrontational nature. It is entirely organized in such a way as to refute the claims of a specific researcher. The criticism of her is personal and direct – she is widely quoted in citations the goal of which is to expose her erroneous interpretation and faulty research approach. Confrontational concessionary structures and quotation marks to express reservations also serve this same purpose. The authors do not refrain from directly undermining the scientific ethos of the researcher they are confronting. The pronouns often hold a dialogue with the reader, but they especially underscore the authors' positions and ethos. All these make the text read like a direct and personal attack and represent a further example of confrontational dialogicity.

5.6 The ping-pong pattern
(Finkelstein 1996; Mazar 1997; Finkelstein 1998)

In this section I will relate to a prime example of what Hunston (2005) calls the "conflict article." This is a case in which a conflict erupts between two researchers that can in actual fact culminate in a succession of three papers: the initiating article, another paper that in its title defines itself as a response to that article, and yet a third article, defined as a response to the response. In the analysis of this case, we can see how the paper's confrontational nature intensifies as the confrontation begins to emerge as a personal one rather than a conflict between scientific claims. This situation constitutes a danger: The discourse community may view a personal conflict as one that is incompatible with the accepted norms of the discipline, that shifts the discussion in a direction that does not contribute anything to the community's shared goals. In such a situation, the community loses interest in the debate and dismisses its importance. In the analysis of the papers in this section, I will try to show how a scientist can contend with this problem. The papers that will be analyzed here are Papers 8, 9 and 10 in Appendix B.

Because the first paper in the series is an initiating paper and the second, which confronts it, resembles the one we saw in the previous section, the current analysis will focus mainly on the third paper in the series. However, in order to understand the content and argumentative context, I will briefly introduce the first two papers as well.

Israel Finkelstein is an archaeologist from Tel Aviv University and is one of the co-authors of the paper analyzed in Section 5.5 above. Amihai Mazar is an archaeologist from the Hebrew University of Jerusalem (and is related to Eilat Mazar; both are related to the late archaeologist Benjamin Mazar: Eilat Mazar is his granddaughter and Amihai Mazar is his nephew).

Finkelstein's position as presented in the 1996 paper (and also published in additional papers and books in Hebrew and English) caused something of an uproar in Israel, one that exceeded the boundaries of the academic ivory tower and spilled over into the media, mainly due to the implications his views had for the nature or perhaps even the very existence of David's and Solomon's United Monarchy in the tenth century BCE. In the absence of any direct archaeological finding from Jerusalem, the assessments regarding this period are based on findings from the areas over which the United Monarchy was supposed to have ranged. These findings are partially consistent with details that appear in the biblical text, and of especial importance is this verse: "And this is the account of the levy which king Solomon raised; to build the house of the Lord, and his own house, and Millo, and the wall of Jerusalem, and Hazor, and Megiddo, and Gezer" (I Kings 9:15).

Hazor, Megiddo and Gezer are three sites that have been excavated by archaeologists and that have produced impressive findings to provide evidence of their economic and political might. These findings have been traditionally dated to the tenth century BCE and are directly related to the above verse from Kings, i.e. to the period of Solomon's reign.

Finkelstein proposed a new dating method for these and other findings, a method that he called "Low Chronology." According to his Low Chronology, the findings that were thought until now to belong to the tenth century BCE are viewed as belonging to a period 50 to 100 years later. What this means is that a strong, large kingdom that constructed monumental buildings and maintained and developed an administrative apparatus did not yet exist in the tenth century, but rather only in a later period, in the time of the dynasty of the northern House of Omri in the ninth-eighth centuries BCE. Finkelstein's acceptance of Low

Chronology significantly weakens the claim that supports the existence of the kingdom of David and Solomon as described in the Bible. As Finkelstein himself writes in the abstract of his paper (Paper 8 in Appendix B):

> (1)   The new dating calls for a re-evaluation of the historical, cultural and political processes that took place in Palestine in the eleventh–ninth centuries BCE.
>
>   (Finkelstein 1996: 177)

At the same time, it should be noted that Finkelstein himself tends to minimize the implications of his study from this perspective. Towards the end of his paper, he writes:

> (2)   Needless to say, all this has nothing to do with the question of the historicity of the United Monarchy. The kingdom of David and Solomon could have been a chiefdom, or an early state, in a stage of a territorial expansion, but with no monumental construction and advanced administration.
>
>   (Finkelstein 1996: 177)

However, contrary to this position by Finkelstein, scholars of the revisionist-minimalist school (whom I mentioned in the analysis of Na'aman's paper in Section 5.4 above and whom I will further discuss in Section 5.7 below) employed his proposal and the conclusions that seemingly arise from it in order to deny the very existence of the United Monarchy.

Finkelstein explicitly denies archaeological research that is based on the biblical text as evidence, as we saw in the previous section in his criticism of Eilat Mazar's approach. The long citations he presents in his paper, especially in the section that describes the history of archaeological research, are from other archaeologists who are in his view similarly misguided. For example, he maintains that the identification of Solomonic Megiddo and Hazor was based solely on the Biblical testimony. Regarding Yigael Yadin, one of Israel's most venerable and prominent archaeologists, he writes that:

> (3)   His understanding of the gates at Meggido, Hazor and Gezer, in accordance with I Kings 9:15, as 'blueprint architecture' of the Solomonic era, became the most important pillar of the archaeology of the United Monarchy. So important that even Aharoni (his great adversary on the stratigraphy of Megiddo, the pillared buildings and other matters) admitted that "this is one of the rare examples in archaeology where the exact date of a building can be determined even without the discovery of any inscription" (Aharoni 1972: 302).
>
>   (Finkelstein 1996: 179)

He further cites a paper by Yadin regarding the contribution of the biblical text:

(4) "Indeed, it seems that there is no example in the history of archaeology where a passage helped so much in identifying and dating structures in several of the most important tells in the Holy Land as has I Kings 9:15" (Yadin 1970).

(Finkelstein 1996:179)

These quotes are of course cited in order to be critical and to claim that in fact "all three foundations of the archaeology of the United Monarchy have been shown to be far from reliable, undisputed chronological anchors."

I will not analyze this paper in detail but only relate to one striking linguistic element and that is the use of the first-person pronouns. In this paper, 18 singular first-person pronouns and 11 first-person plural pronoun forms could be found. I will present the most outstanding and important of them below.

The plural forms are generally inclusive and relate to the research community as a whole, the shared knowledge of its scholars and their shared goal – to seek the truth:

(5) If *we* could safely identify the relevant strata, *we* could be able to produce a vivid picture of the material culture and settlement patterns of the tenth century BCE [...].   (Finkelstein 1996:179)

(6) The historical reliability of I Kings 9:15 has not been confirmed; that is, *we* do not know whether it is Solomonic in date, or whether it describes the supposed events of the tenth century from a later perspective.

(Finkelstein 1996:179)

(7) The first important synchronism is the campaign of Pharaoh Shoshenq to Palestine in 926 BCE. But *we* need to admit that *we* do not have even one destruction layer which can safely be assigned to this campaign.

(Finkelstein 1996:180)

Especially interesting is the first-person pronoun form in the paper's concluding paragraph, in which the author suggests what archaeologists should do in the future:

(8) A crucial challenge for future archaeological research in Iron II sites, especially in the north, will be to reveal new, understandable chronological anchors for the tenth-ninth centuries BCE. Until this is achieved, *we* are left with two alternatives, not being able to give a clear-cut verdict between them. The matter is therefore left to the overall historical and cultural understanding of each scholar.   (Finkelstein 1996:185)

The second plural pronoun emphasizes the author's membership in the discourse community and his dialogue with it, and he thus submits his proposal for the judgment of the other scholars and for their use in order to advance their shared scientific enterprise.

In another example, the inclusive plural form is directed at the readers and its role is to include them in the advancement of the claim:

(9)   But *we* need to undertake this step by step, and start at the beginning, that is, with a short resume of the history of the search for the archaeology of the United Monarchy.   (Finkelstein 1996: 177)

The singular forms, on the other hand, underscore the presence of the researcher in the text and emphasize his place as the author and developer of the argument:

(10)   In the following lines *I* wish to present briefly *my* proposal for a Low Chronology for the Philistine settlement (Finkelstein forthcoming).
                                                       (Finkelstein 1996: 179)

(11)   *I* sincerely believe that if these datings could have been proven beyond doubt, there would have been no difficulty in demonstrating that in the tenth century there was a strong, well-developed and well-organized state stretching over most of the territory of western Palestine.   (Finkelstein 1996: 177)

This emphasis on the presence of the author comes fairly often in order to emphasize his criticism of others, sometimes in rather harsh statements, as in the following examples:

(12)   *I* would take the risk of asserting that by ignoring archaeology the debate has become, in many ways, futile.   (Finkelstein 1996: 177)

(13)   Turning to the archaeological anchors for the early-Iron II, *I* would start by emphasizing that some of them […] are no more than vague archaeo-historical assumptions and there is no need to deal with them here.
                                                       (Finkelstein 1996: 180)

In the following example too, he hints at the drawbacks in the work of others, and on this background, the first-person pronoun emphasizes the preferred approach that he proposes:

(14)   In what follows, *I* wish to discuss the search for the archaeology of the United Monarchy free of any conventional wisdom, text bias, or irrelevant sentimentality.   (Finkelstein 1996: 178)

Nevertheless, the author may also point to himself as one who is willing to admit past mistakes and correct a previous dating of his own:

(15)   Following the generally accepted chronology, *I* dated the site from the second half of the eleventh century to the end of the tenth century BCE [...]. Opting for the Low Chronology, a date in the mid tenth-early ninth century BCE seems a better alternative for this one-phase settlement.

(Finkelstein 1996: 182)

In contrast, in the following example, we can see that he refers to a mistaken approach that in his opinion belongs to the past, using the first-person plural pronoun:

(16)   The principal disadvantage is for Biblical history, or at least for the way *we* used to comprehend it: the alternative scheme puts the monuments previously dated to the second century in the early ninth century BCE.

(Finkelstein 1996: 184)

The inclusive plural form serves here in order to include himself as part of the discourse community and create a partnership between himself and the community of researchers. As noted above (in Section 5.1), a plural form of this kind can serve as a politeness device, which mitigates the criticism expressed by the author. As Myers notes, "the paradox, of course, is that the writers are identifying themselves with the very ideas they are refuting – usually [...] by showing how the field has changed" (Myers 1989: 7).

The singular form also serves Finkelstein to establish his own ethos as an archaeologist who actually "was there" and saw the findings with his own eyes:

(17)   Recently, *I* have checked anew the problem of the date of the settlement of the Philistines in the southern coastal plain of Canaan (Finkelstein forthcoming).

(Finkelstein 1996: 177)

(18)   In an attempt to shed light on the relative chronology of the Beer-sheba basin sites, *I* have checked the main pottery forms typical to the region [...].

(Finkelstein 1996: 181)

I will make do with this in order to present Finkelstein's ideas in this paper, and will now move on to the second paper. As Hunston (2005) argued, any paper has the potential to be an initiating article that leads to confrontational responses. However, some papers have a provocative content that "invites" a direct response. As we have seen, Finkelstein's article is indeed of this nature. His conclusions are far-reaching and their implications for the world of research could be very significant. I will now present Amihai Mazar's response.

Amihai Mazar's article (1997, Paper 9 in Appendix B) is defined as a response to Finkelstein already in its title, and was published in the following issue of the same journal. It in fact refers directly to this paper as well as to another one published in 1995, which I have not analyzed here. Mazar's declared aim is clearly stated in the concluding sentence of the Introduction:

(19)   Recently, however, I. Finkelstein (1995, 1996) has proposed lowering the dates of Iron Age assemblages by some 50–100 years, against the commonly accepted dates. The following short reply seeks to demonstrate that the older scheme is preferable.   (A. Mazar 1997: 157)

The definition of this aim frames the dialogic nature of the paper from the outset: The main dialogue in it is held with the specific proposal put forth by a specific researcher, and not with the ideas of a group of researchers, with consensual knowledge or trends in research. Consequently, I will focus in my analysis on this specific dialogic aspect and keep discussion of other aspects, which have been discussed in previous sections, to a minimum.

Throughout the article, Finkelstein's claims are systematically rejected one by one, and the author makes use of a number of rhetorical tools in order to do so.

He positions Finkelstein's claims as a groundless scientific hypothesis, in contrast to the scientific proof that emerges from the Carbon 14 dating:

(20)   14C testing of some carbonized grain from a small bin at Beth Shean [...], these dates provide the first and only scientifically secure evidence for the eleventh century BCE date of these assemblages, as opposed to the late tenth century BCE date suggested by Finkelstein.   (A. Mazar 1997: 160)

The proposal is presented as lacking foundation also by the mention of findings of the author himself, with his own ethos and thereby also the validity of the findings, are emphasized by means of the use of the first-person pronoun:

(21)   The archaeological data also contradict Finkelstein's suggestion (1996: 183–184) that there was an occupation gap at Beth Shean after the Egyptian presence there. In *our* excavations, *we* observed a considerable amount of continuity in town planning, pottery, and other aspects of material data [...].
                                                                        (A. Mazar 1997: 160)

Mazar points to inconsistency in Finkelstein's method, which ultimately causes a contradiction:

(22)   Finkelstein suggests lowering most of these assemblages to the ninth century, except Arad Stratum XII, which for some reason he accepts as being destroyed by Shoshenq. His reasoning is flimsy: it is based on [...] a limited resemblance of the pottery (in his words: "somewhat similar") at Yezreel to that of Megiddo IVB–VA [...]. Since Finkelstein now wishes to lower the date

of these other assemblages to the ninth century BCE, I would expect that he would also lower the date of Arad XII. By not doing so, he has created an internal contradiction in his reasoning. (A. Mazar 1997: 160–161)

The two-word quote from Finkelstein ("somewhat similar") is particularly interesting here. This quote appears once again immediately afterwards:

(23)   Since it was founded by the Omrides in the ninth century, and its pottery is "somewhat similar" to that of Megiddo IVB–VA, he claims that a ninth century BCE date is indicated for the latter. (A. Mazar 1997: 161)

When the quoted phrase appears a second time, the irony becomes evident, and this ironic tone helps the author to convince the reader of the unreasonableness of Finkelstein's claims. Here is yet another utterance that can be seen as ironic:

(24)   Thus it seems to me that the 'mystery of the missing century' is a mere illusion. (A. Mazar 1997: 163)

The phrase 'mystery of the missing century' is not a quote from Finkelstein's article, although it does echo his ideas. One of Finkelstein's claims in justifying the Low Chronology is the fact that if the traditional High Chronology is used, there would be no findings from the ninth century BCE for numerous sites. The Low Chronology means that findings that were thought to represent the tenth century BCE would now be thought to represent the ninth century, thus resolving the problem. Mazar gives this issue a name and formulates it so as to mark it linguistically: The formulation given within single quotes appears to belong to a different genre, and is perhaps reminiscent of the name of a mystery or adventure book or a newspaper headline. The echoing of Finkelstein's ideas in this way is an ironic echoing that underscores Mazar's reservations with them and places them in a ridiculous light. Perhaps the author is also alluding to the popularization of these ideas and the interest they awaken among the public.

In other places, Mazar points to the improbability of Finkelstein's claims, which lead to illogical conclusions:

(25)   As for Lachish, Finkelstein (1996: 181) relies on a supposed similarity [...] to claim a ninth-century date for this level [...]. This would leave Lachish unsettled during the tenth century BCE, an unfeasible conclusion concerning this major Judean site. (A. Mazar 1997: 161)

(26)   The lowering of Stratum X at Hazor to the ninth century BCE as suggested by Finkelstein creates an impossibly brief duration for each of these strata. (A. Mazar 1997: 163)

Another way to refute Finkelstein's claims is by mentioning sites that he omitted from his discussion:

(27) The pottery assemblage of Kuntillet 'Ajrud (Ayalon 1995) not mentioned by Finkelstein, is of particular importance for our subject. [...] At Taanach, another site not mentioned by Finkelstein, a fine tenth-century BCE assemblage was studied by Rast (1978), who based his conclusions on careful comparative study of "Primary Comparative Loci" at other sites.

(A. Mazar 1997: 162)

The mention of the importance of these two sites in the discussion, and the findings from the site whose dating is based on a "careful comparative study" imply that these two sites were not mentioned by Finkelstein in his paper perhaps because doing so could weaken his case.

In addition, a common scientific claim is used to argue against Finkelstein's claims: that his proposal, which aims to solve problems in research, creates new ones of its own:

(28) Finkelstein's ideas concerning Megiddo (1996: 183) raise formidable questions, rather than solving problems. (A. Mazar 1997: 161)

In this context, Mazar's description is interesting:

(29) The proposal to lower the date of Philistine Monochrome pottery has created a 'snowball' effect, for now Finkelstein has to date the appearance of Bichrome Philistine pottery to the beginning of the eleventh century BCE.

(A. Mazar 1997: 159)

The expression 'snowball effect' relates to the undesirable results of accepting Finkelstein's proposal, which Mazar continues to describe in detail further on. The negative connotation of the expression relates to the aspiration for coherence in research. A new proposal, one that cannot integrate coherently into the claims that have been accepted by the discourse community and consequently requires it to reconsider them is one that is more difficult to accept. From this it follows that there has to be a very good reason to justify this change. This point is brought home in Mazar's text, for example when he reiterates his claims a number of times that there is no reason that could justify the re-dating that Finkelstein suggests, for example:

(30) I see no reason for such a lowering of dates. (A. Mazar 1997: 160)

(31) If such was the case in the Late Bronze Age, I see no reason not to accept a similar situation in the Iron Age. (A. Mazar 1997: 161)

(32)   I see no difficulty in assuming that the monumental buildings of Stratum IVA
        were founded during the mid-ninth century BCE […].

                                                                (A. Mazar 1997: 161)

(33)   I see no difficulty in retaining the "Solomonic" date of the monumental ashlar
        buildings […].   (A. Mazar 1997: 164)

He offers his conclusion of this matter at the beginning of the Conclusions section
of the paper:

(34)   In my view, there is no justification to lower by 50–100 years the traditional
        dates […].   (A. Mazar 1997: 164)

The scientific logic that underlies this statement is that a significant claim such
as this cannot be accepted if there is no real justification for it, because it exacts
a high "price": It creates new problems and would require extensive changes in
the body of accepted knowledge. An argument of this kind appeals directly to the
discourse community, its conventions and its aspirations to defend the coherence
of its shared body of knowledge.

    As for the manner in which the citations are presented, Finkelstein's claims
are generally presented without quotation marks and are formulated in Mazar's
words. Consequently, the direct quotes, when Finkelstein's actual words are cited
are especially noteworthy. Of special importance is the place where Mazar cites
from Finkelstein's conclusions, right at the end of his article:

(35)   One of the main conclusions of Finkelstein's study is his statement that "from
        the archaeological perspective, the line between the Iron Age I and the Iron
        Age II… should be put in the early ninth century rather than c. 1000 BCE."
        In the following paragraph he claims that "all this has nothing to do with the
        question of the historicity of the United Monarchy" (Finkelstein 1996: 185). I
        find myself in disagreement with both statements.   (A. Mazar 1997: 163)

It is easy to understand why these quotes were cited verbatim. The first is in fact
Finkelstein's main conclusion and the second is a statement that seeks to narrow
its implications. The first is a classic archaeological argument and the second is its
possible use for the purpose of the historical discussion. The direct quote helps
Mazar to drive home his two points: First, he is opposed to the new dating based
on archaeological arguments, and second, he believes that accepting the new dating would necessarily lead to a change in the historical perception of the United
Monarchy period, something he has difficulty accepting.

    It is interesting to note that in all the later Examples (30)–(35), the author
uses the first-person singular pronoun. In this article, Mazar makes considerable

use of these forms to underscore his own positions and conclusions. Thus, in the paper's two concluding sentences:

(36)   As long as these two are not available, *I* prefer the traditional chronology. In *my* view, Finkelstein's conclusions concerning the archaeological background of the United Monarchy are premature and unacceptable.

(A. Mazar 1997: 165)

In this paper, the author did not refer to himself using the plural first-person pronoun *we*, but on the other hand, he twice refers to himself as 'the author'. He uses the plural form for two separate purposes. One is to refer to himself as someone who excavated in Beit Shean. In this case, the exclusive plural is an accepted form to mention the group of individuals who participated in the dig:

(37)   [...] since in *our* excavations at Beth Shean, imported sherds of Myc. IIIC were found [...].   (A. Mazar 1997: 159)

(38)   At Beth Shean, *we* uncovered fragments of monumental structures [...].

(A. Mazar 1997: 164)

In other cases, he uses the inclusive plural form to hold a dialogue with the discourse community. In doing so, the author refers to the shared knowledge and goals of the community, for example:

(39)   During this time *we* have no sound basis for absolute chronology [...].

(A. Mazar 1997: 157)

(40)   Yet if *we* maintain the higher date for the Monochrome phase, there is no reason to lower the date of the Bichrome phase.   (A. Mazar 1997: 159)

(41)   This radical suggestion, if corrected and accepted, would have far-reaching implications for *our* understanding of the archaeology and history of the tenth/ninth centuries BCE in Israel [...].   (A. Mazar 1997: 159)

(42)   Moreover, *we* should not expect great changes between the pottery of the second half of the tenth century [...].   (A. Mazar 1997: 161)

(43)   This assemblage demonstrates that at a single point in time *we* may find a combination of older and more recent forms [...].   (A. Mazar 1997: 162)

(44)   In many sites *we* can see similarities in the pottery during several stratigraphic phases [...].   (A. Mazar 1997: 162)

Mazar's affinity to the research community is especially evident when he discusses the difficulties and limitations that all the members of the community share, for example:

(45) Another problem is *our* methodology in discussing pottery: in most cases *we* rely only on the appearance or disappearance of diagnostic types.

(A. Mazar 1997: 162)

(46) Even then, it might be that *our* ability to use pottery for 'fine tuned' dating will remain limited.   (A. Mazar 1997: 162)

The reference to the various elements of the discourse community – its shared knowledge, research and deduction methods, its goals and the problems that it faces – strengthens the affinity between the writer and the research community, presenting him as a valuable member of it.

One example of the use of the plural form appears to be related to the process of the argument and how it is presented, and thus holds a dialogue with the reader:

(47) This brings *us* to the more general question of the chronological span [...].

(A. Mazar 1997: 162)

Amihai Mazar makes considerable use of concessionary structures to hold a dialogue with Finkelstein and reject his claims. The concessionary structures make it possible to accept some of Finkelstein's claims, but to reject the conclusion drawn from them. Thus, for example, in the following example, there is a certain acceptance of the claims in the satellite position, but in the nucleus position, the implied conclusion – the Low Chronology – is rejected:

(48) The latter two claims are to a certain extent true, *yet* both can be satisfactorily explained without a wholesale lowering of the Iron Age chronology of Israel.

(A. Mazar 1997: 160)

The following paragraph contains three concessionary structures, in which Finkelstein's claims are referred to briefly in the satellite part.

(49) *Despite* the differences between the imported Myc. IIIC and Philistine Monochrome, they are still very close to one another [...]. *Though* not decisive, the stratigraphic evidence from Beth Shean [...] suggests a time before the end of Egyptian domination in the country. Thus, *in spite* of the difficulties pointed out by Finkelstein, I see no reason to push the beginning of the Philistine settlement [...].   (A. Mazar 1997: 159)

A further rhetorical tool that Mazar uses is the rhetorical question, for example:

(50) His suggestion is based on the lack of Philistine pottery at Beth Shean. Yet why should we expect to find Philistine pottery at Beth Shean? The site is far away from Philistia [...].   (A. Mazar 1997: 160)

The rhetorical question aimed at the audience – the reader and the entire discourse community – is answered in the text and serves as a persuasive element. A further rhetorical question stands out in the text as relating not to the interpretation of the archaeological findings but rather to the broader dispute being held over the history of the area in the tenth-ninth centuries BCE:

(51)   Finkelstein's 1996 paper would encourage historians who tend to the minimalistic or even the nihilistic approach in evaluating this period. Was Finkelstein's interpretation of archaeological data inspired by some current trends in historical writing?   (A. Mazar 1997:164)

The rhetorical question takes on an accusatory tone here, and can be seen as an undermining of Finkelstein's ethos as an objective scientist. However, as Mazar accuses Finkelstein of an ideological bias, the first sentence in Example (51) reveals the ideology that guides Mazar himself when he expresses his concern in view of the possibility that Finkelstein's proposal might serve to bolster the revisionists' claims. The text of this article reveals possible ideological aspects in other points too, for example, the name given to the geographic area under discussion: Whereas Finkelstein, for example, calls it "Palestine," Mazar calls it "Israel" or "Eretz Israel," as it is known today, the time of the modern State of Israel, although he means to include in it both the regions of Judea and northern Israel, and in some cases also Philistia (the southern coastal plain) and Phoenicia (the northern coast). This is demonstrated in the following examples:

(52)   Two well-dated events currently anchor Iron Age chronology in *the land of Israel* at both ends of the historical spectrum [...].   (Mazar 1997:157)

(53)   This effect has remained almost unnoticed in pottery studies in *Israel* [...].
                                                                (Mazar 1997:162)

(54)   Clearly defined borders between two neighboring cultures [...] can be seen in various periods, even in a small country like *Israel*.   (Mazar 1997:158)

(55)   The appearance in Philistine urban centers like Ashdod and Ekron [...] is a very unusual phenomenon in the archaeology of *Israel*. This pottery represents one of the few cases in the archaeology of *our* region [...].
                                                                (Mazar 1997:158)

The use of the first-person plural pronouns in the last example is yet another example of a seemingly "ideological" and non-scientific aspect revealed in Mazar's writing. These linguistic choices, which are apparently perfectly natural for him, just as they are for many Israeli researchers, perhaps reflect the sense of belonging on the part of the contemporary Israeli researcher who lives in the State of Israel of today, who often views it as a continuation of the ancient Judean kingdom.

Israeli archaeologists often need to contend with "suspicions" regarding their ability to interpret the findings objectively.

In addition, Mazar's tendency to accept the biblical story at face value, at least partially, emerges from his writing. This tendency is evident for example in the excerpt here, with the biblical reference at the end.

(56)   The latter date is based on the assignment of these destructions to the conquests of David in the coastal plain and northern valleys, which can tentatively be dated between 1000–980 BCE (Mazar 1993: 223; 1994: 49). Although such military conquests are not mentioned in the Bible, they probably occurred, since these regions are known to be part of the Israelite Kingdom during the time of Solomon (I Kings 4: 11–12).   (A. Mazar 1997: 160)

In the analysis of Finkelstein's response article (to follow), we will see how all these elements are used against him.

In summary, this article belongs to the confrontational pattern. It has a clearly defined aim: to refute the claims of a researcher in a specific article, and to that end, various dialogic elements are employed: the confrontational concessionary structures, rhetorical questions, citations, some of which are used to create an ironic tone. The first-person pronouns underscore the author's authority, on the one hand, and hold a dialogue with both the discourse community and the reader, on the other.

Finkelstein's second article (1998, Paper 10 in Appendix B) is defined as a response to a response. It is not a full-length article and the author does not take the trouble to present the full picture. He turns directly to Mazar's claims and makes very little mention of other researchers (although his bibliography contains 35 items, of which three are his and two are Mazar's). The main thrust of his argument is that the disagreements between him and Mazar are not over archaeological findings, but are rather the result of their different research approach, which affects the way each of them interprets the same findings. The title of the article reflects his desire to move the debate in this direction: "Bible Archaeology or Archaeology of Palestine in the Iron Age?" The wording of the title as a question is important because of its rhetorical effect. The disjunction creates a fallacy called "poisoning the wells," in which the opponent gets the "choice" of drinking at one of two wells after the arguer has unfairly tainted one of them (Weddle 1978: 20).

The disjunction implies that only one of the two possibilities can exist, and that the second alternative should be rejected. Since the phrase "Archaeology of Palestine in the Iron Age" is clear and neutral, i.e. there is no reason not to accept it, this leaves the phrase "Bible Archaeology" as a problematic term and an option that should be rejected. In a footnote, Finkelstein explains what he considers

"Bible Archaeology" to be: "archaeology which is dictated by uncritical reading of the Bible." The negative value of the adjective 'uncritical' in a scientific context makes it clear that this method of archaeological research cannot be considered the preferred one. The title of the article, which is worded in the form of a question, can therefore be interpreted indirectly as a claim for the preference of the view to be presented by Finkelstein. Incidentally, it is notable that despite Finkelstein's declared opposition to any reliance on the Bible in archaeological research, he himself does not refrain in his own research from associating his archaeological findings with the biblical text.

The question of ideological bias also comes up in the article's opening paragraph, which includes two quotations from Mazar's article, the second of which is that same accusatory rhetorical question that Mazar directed at Finkelstein (see Example (51) above):

(57) Mazar (1997) argues that my Low Chronology for the early Iron II strata (Finkelstein 1996a) "leads to a *distorted* [*my emphasis – I.F.*] archaeological picture of the period of the United Monarchy". Mazar, then, takes the position of the custodian of the ideal, harmonic picture of Bible archaeology, which argues for a glorious Solomonic state, against an intruder, who threatens to shatter the sentimental images. In the closing remarks of his rejoinder, Mazar points to a possible motive behind the onslaught, asking, "Was Finkelstein's interpretation of archaeological data inspired by some [*unnamed – I.F.*] current trends in historical writing?" The truth is that this is not a fashion or a fad; rather, it is an effort to scrutinize the distorted roots of Bible archaeology and to establish the archaeology of the Iron II on independent foundations.

(Finkelstein 1998: 167)

The struggle over the scientific ethos of the two researchers, the harbingers of which could be seen in Mazar's article, is strongly underscored here in the article's opening paragraph. The paragraph's linguistic design is interesting from a number of aspects. In the first citation from Mazar's article, Finkelstein emphasizes the adjective 'distorted', using italics, and then paints a harsh picture of a heroic struggle between good and evil: on one side are the defenders of Bible Archaeology who grasps onto ideal, harmonic and sentimental images, and on the other side is the intruder, who is defined as a threat. Finkelstein's opinion on who the "good guys" and "bad guys" are in this struggle becomes clearer after the second quote. Finkelstein once again uses the adjective 'distorted', which is taken from Mazar's words." However, this time, he uses it in order to argue that the approach that ought to be described by this adjective is not his own, but rather the opposite one, the one that he calls 'Bible Archaeology'. He presents his own approach as one that aspires to independent foundations – in this context, the adjective 'independent' refers to objectivity and being free of mistaken preconceptions.

Mazar's rhetorical question is cited in order to provide it with a very different response from the one that Mazar hints at in his article. The term 'trends' has a negative value in this context in that a researcher who is influenced by trends may arrive at biased conclusions. Trends also have a temporary and transient nature, and consequently, a researcher who is influenced by them at any stage of this research may then change his positions after the trend has passed. Finkelstein dismisses this possibility in his response by explicitly stating: "This is not a fashion or a fad." From this it follows that he himself and his approach are presented here as the scientific, objective "good" that is not affected by fashion and that will liberate the archaeological world from previously held tendentious perceptions.

Throughout the article, Finkelstein rejects Mazar's claims in detail, and the text is arranged in accordance with the various excavation sites. To reject the claims, numerous quotations from Mazar's text are cited.

Two of the quotations are cited in order to demonstrate that Mazar's logic is circular. Regarding the site known as Taanach, Finkelstein argues that:

(58)   Arguing that the pottery assemblage from period IIB "is well dated to the second half of the tenth century" (Mazar 1997: 160) is circular reasoning.
                                                                              (Finkelstein 1998: 171)

At this point in the text, the reader's attention is directed to a footnote, where Finkelstein points to yet another example of circular logic:

(59)   For similar circular arguments see the discussion of the 'Hippo' jars in Mazar (1997: 160).   (Finkelstein 1998: 173)

In another place, Mazar is attacked along with another researcher, and yet a third is noted, to demonstrate yet a third example of circular reasoning. Regarding the site known as Kunntillet 'Ajrud, Finkelstein writes that:

(60)   This site [...] exemplifies the breaches in Mazar's method. Citing Ayalon's study [...], Mazar argues that 'Ajrud is "one of the few relatively secure anchors for pottery chronology in the period under discussion" (1997: 160). Ayalon is cautious, maintaining that "it appears that the ceramic assemblage of 'Ajrud must be dated to the time span between the end of the ninth century to the beginning of the eighth century B.C.E. This conclusion is corroborated by Carbon-14 dating of samples of 'Ajrud." However, the 14C dates from 'Ajrud can be interpreted in more than one way, and Segal (1995) admits that a mid-eighth-century date is an alternative. Reading Segal's report, there is no doubt that the final interpretation of the results was influenced by the

historical and archaeological dating of the site – another example of circular reasoning.   (Finkelstein 1998: 171)

This paragraph merges the views of a number of researchers and its dialogicity can be represented schematically as follows:

Finkelstein	↔ against ↔	**Mazar**
supports		supports
Segal	↔ against ↔	**Ayalon**

Mazar enlists Ayalon to support him. In order to contend with that, Finkelstein enlists Segal, whose claims weaken Ayalon's arguments. In this way, he manages to indirectly undercut Mazar's claims. In other words, he directly cites both Mazar and Ayalon for the purpose of refutation. Further direct quotations are cited from Mazar in order to either explicitly or implicitly reject his claims:

(61)   A regional solution ("the site is far away from Philistia" – Mazar 1997: 158) is impossible in this case […].   (Finkelstein 1998: 170)

(62)   Mazar argues that the lowering of Hazor X to the ninth century "creates an impossibly brief duration" for Strata X–V (1997: 161). In the case of Hazor, we must wait for the publication of the results of the renewed excavations.
                                                                    (Finkelstein 1998: 171)

Besides Finkelstein's dialogue with Mazar and other specifically named researchers, he also holds a dialogue with the discourse community by referencing shared knowledge that is used to reject Mazar's claims:

(63)   However, *there is little doubt today* that the Khirbat Karak phase did not cover the entire span of Early Bronze III.   (Finkelstein 1998: 168)

(64)   First, the Bichrome pottery *is known* to have been popular mainly on the coast […].   (Finkelstein 1998: 168)

A further dialogue that Finkelstein holds in his article is a direct dialogue with the reader through the use of the term 'the reader' and obligation modals:

(65)   *The reader should* be reminded that we are dealing with sites located only several kilometers away from each other […].   (Finkelstein 1998: 168)

(66)   […] *the reader should* be aware of two more issues: […].   (Finkelstein 1998: 169)

A number of references to Mazar's claims are presented in a concession structure. Finkelstein explicitly expresses agreement with Mazar in the satellite of the concession, only to reject his entire claim immediately afterwards:

(67)   Mazar (1997:159) is right in dating Arad XII to the tenth century. The con-
temporaneity of this stratum with Lachish V is also possible. *But* the fact
remains that the first fortified sites in the south [...] post-date the tenth century.   (Finkelstein 1998:171)

(68)   I agree with Mazar that a sufficient time span should be allowed to separate
the 'Ajrud corpus and my ninth-century assemblage. *Therefore*, the date of the
former should be lowered to the first half, or mid-eighth century B.C.E.
                                                          (Finkelstein 1998:171)

(69)   Mazar is right in his statement that the Low Chronology will force me to
change many of my views on the archaeology of proto-Israel, a fact which
I acknowledged in my 1996 *Levant* article. His assertion (1997:161) that a
300-year time span is too long for the Iron I phenomenon in the hill country
is less convincing.   (Finkelstein 1998:171)

These three examples clearly demonstrate the fact that the agreement expressed
in the concession satellite does not weaken the claim, because the argumentative
direction of the claim as a whole is determined by the nucleus. In the last example,
the claim in the satellite is presented as lacking any argumentative value even
more explicitly by the fact that the author claims that this is his own argument,
one that already appeared in his previous article.

The strategy of labeling, which we saw Mazar use in the case of "the mystery
of the missing century" (see Example (24) above) also serves Finkelstein in this
article. He relates to the fact that no Egyptian ceramics were found in Philistine
sites, and that no Philistine ceramics were found in sites under Egyptian influence, despite their geographic proximity. Finkelstein explains this by arguing that
they were not there during the same period. Mazar's explanation involves the existence of geopolitical borders, beyond which ceramic utensils did not move. This
is Finkelstein's response to Mazar's suggestion:

(70)   Mazar cites two examples to support his 'apartheid theory'; both are irrelevant.   (Finkelstein 1998:168)

The word 'apartheid' of course does not appear in Mazar's article, where he discusses the phenomenon in the following words:

(71)   Clearly defined borders between two neighboring cultures due to ethnic,
economic, or even ideological differences [...].   (Mazar 1997:158)

Finkelstein consequently uses the word 'apartheid' as an ironic label for Mazar's
position, thereby expressing his reservation with the plausibility of the claim.

Finkelstein further uses additional types of arguments. For example, in a number of places, he maintains that Mazar did not take all the evidence into account:

(72)   The evidence comes from a large number of cites, many more than Lachish and Tel Sera, mentioned by Mazar.   (Finkelstein 1998: 167)

In the following example, he explicitly argues that the omission by Mazar was deliberate, a serious accusation in a scientific context:

(73)   Mazar prefers not to deal with a key question regarding the Hazor chronology – the date of Stratum XII.   (Finkelstein 1998: 171)

Another argument Finkelstein uses is his claim that there is a contradiction in Mazar:

(74)   In sustaining the *dream* of a Great Solomon, Mazar is trapped into a chronological contradiction.   (Finkelstein 1998: 170)

The word 'dream' points to an emotional approach that has no place in scientific research. His accusation that Mazar mixes his feelings into the research comes up again even more explicitly in the Discussion section of the article. Finkelstein uses the adjectives 'sentimental' and 'romantic', which have a negative value in a scientific context because they are perceived as running counter to "scientificity":

(75)   The main obstacle which distracts Mazar from viewing the archaeological data on their own terms is his *sentimental*, somewhat *romantic* approach to the archaeology of the Iron Age.   (Finkelstein 1998: 172)

In order to prove this, Finkelstein cites long quotes from Mazar's article, commenting harshly on them, for example:

(76)   Mazar considers this idea a blasphemy […].   (Finkelstein 1998: 172)

The discussion of Mazar's emotions develops into an *ad hominem* attack aimed at denying Mazar's authority to offer a rational scientific opinion. This continues further on as well:

(77)   In the same line of thinking, Mazar is *frustrated* by the meager tenth/ninth-century finds in Jerusalem […].   (Finkelstein 1998: 172)

The discussion section includes further long quotations from Mazar's article whose aim is to prove that Mazar reads the words of the biblical historian as straight-forward historical testimony. He especially attacks him for the statement I quoted above in Example (56), in which he "uses sources, which are yet to be proven historical, to reconstruct an assumption not supported by any text, even the Bible" (Finkelstein 1998: 172).

An interesting quotation of a word that Mazar uses appears right at the beginning of the section, and is included within a concessionary structure:

(78)   Even if my arguments for lowering the dates of strata commonly believed
       to date to the tenth century might be considered, by some, "flimsy" (Mazar
       1997: 158), they are far more solid than the *shaky* foundations of the prevail-
       ing chronology, which Mazar vigorously defends.   (Finkelstein 1998: 172)

It becomes evident from the concessionary structure that Finkelstein is willing
to accept the unflattering adjective Mazar attributes to his claims, because this
adjective is not strong enough to overcome the adjective 'shaky' that he himself
attaches to Mazar's arguments.

Despite the direct quotation from Mazar, he appears in this quote as part of
a group of researchers, through the use of the phrase 'by some.' Later in the same
paragraph too, Finkelstein continues to relate to other researchers who share similar views to those of Mazar, and in one of the rare cases in this article, actually
mentions them by name:

(79)   However, many centre-stage (that is, not only revisionists) authorities argue
       that the stories of David and Solomon draw a picture of a golden age, rather
       than a fully historical period […] (e.g. Van Seters 1983: 307–312; Garbini
       1988: 32; Auld 1996; Niemann 1997).   (Finkelstein 1998: 172)

According to this, it seems apparent that Finkelstein seeks, at least to some extent,
to present the conflict not as being between two different researchers, but rather
between two different positions. This intention is not entirely consistent with the
fact that throughout the entire paper, he employs a large number of first-person
singular pronouns that underscore his own presence in the text, as can be seen in
Examples (68), (69) and (78) above. His presence is particularly salient in the following example, which not only contains first-person singular pronouns but also
mentions Finkelstein himself as an eye-witness to the findings:

(80)   Professor Gitin kindly presented to *me* the new evidence from the 1995–1996
       seasons at the site, mentioned by Mazar (1997: 163, n. 3). It does not change
       the picture that *I* presented in *my* Tel Aviv article […].
                                                            (Finkelstein 1998: 168)

At the same time, in the article's concluding section, he emphatically and explicitly reiterates that the dispute between himself and Mazar is not a personal one, but
is in fact a difference of opinion between two approaches towards archaeological
research, and that deciding between them lies at the heart of the discipline:

(81)   The dispute over the archaeology of the United Monarchy is a clash between
       two approaches to Iron Age archaeology. The first attempts to deal indepen-
       dently with the archaeological data. The second takes the Bible as the frame-
       work and adjusts the archaeological record to pseudo-historical sources. From

> the point of view of the integrity of *our* discipline, the latter is a disturbing phenomenon regardless of who is right and who is wrong chronologically.
>
> (Finkelstein 1998: 173)

Here too, Finkelstein uses expressions that shed a negative light on the approach represented by Mazar: The term "pseudo-historical sources" notes them as being unreliable, and the act of "adjusting" the archaeological record hints at distorted and biased work methods. All these are described by means of the negative adjective "disturbing." However, of particular interest is the final sentence, in which Finkelstein takes the broader perspective of one who is seemingly uninvolved in the dispute. He underscores his own membership in the discourse community and his concern for the success of its endeavor in general terms, for example by means of the use of the inclusive first-person pronoun 'our', which refers to the community as a whole.

In the final part of the paragraph just quoted – "regardless of who is right and who is wrong chronologically" – Finkelstein engages in an especially interesting strategy from a rhetorical standpoint: He appeals to the "objective reader," as defined by Perelman and Olbrechts-Tyteca (see Section 2.4 above). Perelman and Olbrechts-Tyteca (1958: 60) proposed understanding the concept of objectivity as "the viewpoint of a wider group embracing both the opponent and the 'neutral.' The latter is qualified to judge not because he is neutral […] but because he is impartial: being *impartial* is not being *objective*; it consists of belonging to the same group as those one is judging, without having previously decided in favor of any of them."

According to Perelman and Olbrechts-Tyteca, the person who is objective, neutral and impartial is the one that belongs to that same discourse community, but his position is not predetermined, and he is consequently open to accepting or rejecting the view of the writer. When Finkelstein suggests that the reader can express an opinion in the dispute without deciding who is right or wrong chronologically, he is therefore appealing to the objectivity of the reader as a member of the broader group of members of the discourse community.

However, Perelman and Olbrechts-Tyteca further maintain that not only can an outside observer be perceived as objective, but a participant in the argument can acquire this position for himself as well. To be a neutral and impartial participant in the events, according to Perelman and Olbrechts-Tyteca, means to create "a balance of forces, maximum attention to the interests at issue, but with this attention equally divided among the different points of view" (ibid.).

Finkelstein manages to acquire this position for himself by means of the paragraph that follows immediately afterwards, which is the article's concluding paragraph:

(82)   The dispute will probably be decided by a combination of short- and long-
       term field and laboratory research. The short term includes the results of the
       excavation of Stratum VIA at Megiddo [...]. The long term includes the accu-
       mulation of 14C results and quantitative analysis of pottery assemblages.
       (Finkelstein 1998: 173)

With these concluding words, Finkelstein proposes how the dispute will ultimately be decided: through the use of objective scientific tools. With this statement, he
places himself above the fray and enables the desired norms of the discipline to
speak for him. At this stage, he does not insist on the primacy of his own position,
but rather grants each of the positions an equal opportunity. This type of argument is presented at the end of the article from 1996, which I analyzed earlier:

(83)   [...] we are left with two alternatives, not being able to give a clear-cut verdict
       between them.   (Finkelstein 1996: 185)

By using this method of presentation, Finkelstein clearly occupies the "neutral"
position, as defined by Perelman and Olbrechts-Tyteca. Presenting himself as a
neutral makes a decisive contribution to establishing his ethos as an objective
scientist. As far as Finkelstein is concerned, the discussion between himself and
Mazar will be decided not by means of the findings and their interpretation, but
rather by means of his more strongly established scientific ethos.

Let us try now to characterize the ping-pong pattern from its dialogic perspective. The first article in the series (Finkelstein 1996) is a paper of the classic
pattern. The response article (Amihai Mazar 1997) is an article of the confrontation pattern, which is characterized by an expansion of direct dialogicity with a
specific researcher in contrast to meager reference to the discourse community. At
this stage, by means of the adversarial response, the two researchers are defined as
direct rivals, and the singular characteristics of the third article (Finkelstein 1998)
should be viewed with this in mind. The unique aspect of this article is expressed
in the balance between these two forces. On the one hand, the quotations are
cited almost exclusively from the rival's article (Mazar 1997), and every available
rhetorical device is marshaled to reject all his claims, including pointing to flaws
and contradictions in his arguments, irony and a direct attack on his ethos as a
researcher. On the other hand, focusing on a single rival poses the danger of a
loss of interest on the part of the audience, because a personal rivalry contributes
nothing per se to the advancement of the scientific endeavor. Consequently, the
researcher must appeal to the scientific community in a different way. Finkelstein
does this by positioning the dispute within a much wider context, as a confrontation between different approaches or schools of thought. In this broader context,
the author occupies an entirely different role. After pointing to all the possible

flaws in his rival's work, he halts his direct attack on him, and takes a completely different tack – presenting himself as an objective observer, leaving the responsibility for deciding on the dispute between the two rivals to the research community as a whole.

5.7 Face-to-face interaction

In 1997, the editor of *Biblical Archaeology Review* invited four researchers to participate in a symposium which he called "Biblical minimalists meet their challengers" (Item 11 in Appendix B, which for the purpose of reference, I will call "Panel"). The background for this symposium was the dispute between the minimalist Bible scholars (also known as "revisionists") and their critics, especially archaeologists. The editor's aim was to present the different parties' positions on the main issues under dispute, especially the question of the existence of the United Monarchy of David and Solomon. The situation was defined a priori as a confrontational one through the choice of the participants in the panel, not unlike television debates in which speakers who are known to hold well-defined opposing views are invited to spar in the television studio.

The researchers representing the revisionist position were Niels Peter Lemche and Thomas Thompson, both Biblical historians from the University of Copenhagen. The other two were archaeologist William Dever of the University of Arizona and the Biblical scholar P. Kyle McCarter of Johns Hopkins University in Baltimore.

The revisionists are called "Biblical minimalists," because they "find less in the Bible" (Panel: 28) and believe there is only small and fragmentary early material in it (Thompson, Panel: 32). The editor, who is the moderator of the debate, tries to give the revisionists an opportunity to put their arguments forward for the benefit of the readers of the publication, and he demands that they make clear statements regarding the issues under debate. From their remarks, it becomes evident that the revisionists believe that the biblical text is only a collection of traditions (Thompson, Panel: 32) and that biblical literature presents us only with an image of the past, and reflects the time in which it was completed (Lemche, Panel: 28–29). They deny the existence of a united monarchy led by David and Solomon in the tenth century BCE. They argue that Jerusalem was not settled at all until about 900, that "we don't have a tenth-century Jerusalem" (Thompson, Panel: 34) and that the Biblical David did not exist (Lemche, Panel: 40).

Certain typical linguistic characteristics of this text are the result of the fact that this is a face-to-face interaction, and the editor decided to leave some of them in the printed version too, for example utterances that are cut short because another speaker interrupted:

(1) Lemche: [...] I mean if we had known one of them, and the second was
                unknown to us...
     McCarter: Wait a minute, explain...   (Panel: 37)

The direct dialogicity is also constructed by the use of first names (Example (8) below), the colloquial form of speech and the addressing of direct questions to the interlocutor. For example, when Lemche suggests that the inscription containing the words *Beit David* [House of David], which was found at one of the excavation sites, might be a forgery, Dever is quoted as saying:

(2) Oh geez, come on fellas, I've handled it. Have you?   (Panel: 36)

One of the most important features of face-to-face interaction is the considerable use of pronouns, including the first-person singular pronoun and the second-person pronoun. In the text of the spoken symposium, the number of pronouns is significantly greater than that found in the written articles that I have analyzed in this chapter. Let's look at an excerpt of Dever's remarks, which mixes the use of the first-person singular pronoun, the first-person plural pronoun and the second-person pronoun. Lemche counters by saying that if he does not believe in the existence of King Solomon as a historical figure, then the use of the term 'Solomonic' in reference to the gates in Gezer is erroneous. Dever responds as follows:

(3) *You* know that's not what *I* mean. That was imprecise language on *my* part.
     But *we* all do that. *We* all, in archaeology, talk about the such-and-such palace.
     As an archaeologist, *I* am not looking for Solomon, the Solomon of the Bible.
     I do, however, think *we* will end up being able to show that state-formation
     processes begin in Palestine in the tenth-century B.C.E. And if *we* can then
     correlate [this evidence] with some reading of the text, *I'm* not opposed to
     that. Are *you*? If the texts and the artifacts seem to converge, is there a prob-
     lem?   (Dever, Panel: 41–42)

In this excerpt, Dever uses two questions that he addresses to Thompson. Seemingly, although Thompson can and should respond, he does not. A second look shows that these are rhetorical questions since Dever is asking Thompson to agree to something that is clearly unreasonable for him to agree to. A positive response to the question could present Thompson as someone who is prejudiced by his ideology.

The pronouns 'my', 'I' and 'you' in this excerpt refer to the two speakers and are directly derived from the immediate interaction. *You* in the question that is aimed directly at the interlocutor especially reflects this. This can be seen in the previous example too.

The *we* forms, on the other hand, deviate from the direct context of the interaction. The two first ones include the members of Dever's discourse community – the archeologists. They can be viewed as exclusive pronouns through which Dever distinguishes himself from his interlocutors, Lemche and Thompson, who are not archaeologists and consequently may not be familiar with the conventions of the discipline. The two final appearances of *we* in this example somewhat ambiguously refer to one of two possibilities: *we* in these utterances could exclusively represent the group of archeologists, or even a smaller group – the archaeologists who belong to the same school as Dever; it could, on the other hand, also refer inclusively to the group of all scientists who seek the truth.

We forms can also be derived from the interaction and include only the two interlocutors in the specific context. For example, at a certain point in the debate, there is a head-on confrontation between Dever and Thompson, one that turns out to have a long history. It begins with an attempt to sketch out the boundaries of the consensus between them:

(4) Dever: [...] Tom, here's where the anger between *you* and *me* comes to the surface too easily, and *I* apologize if it does: *We* are trying to do the same thing, but *we're* badly misunderstanding each other. *We* need to get rid of the polemics and get to the issues. *We* do agree on a lot of things. *I'm* not nearly as *radical* as *you* guys make *me* out, and *you're* probably not as *negative* as *I* think you *are*. Here's something on which *we* might agree: Archaeology can provide a different kind of context. It's not just a context of literary transmission.

Thompson: Absolutely.

Dever: Now, let's go back to Solomon because that's where *we* really begin to get in trouble. *We* would agree on a lot of things about the patriarchal era and the so-called conquest era, but when *we* come to the monarchy, *I* think, that's where the differences between *us* become plain. Don't *you*?

Thompson: If *you* want to focus on where *we* have disagreements, *I* think that's quite correct.   (Panel: 33)

The adjectives 'radical' and 'negative', which are judgmental, are used here by Dever as descriptions that he assumes will turn out to be inappropriate. Despite the fact that, it is clear from the outset that there is a serious, perhaps even hostile dispute raging between these two researchers, at this stage, Dever tries to narrow the conflict and emphasize those things upon which they can agree. By use of the terms "the patriarchal era and the so-called conquest era," he points to the issues that most current researchers can agree upon: Neither the stories about the Patriarchs (Abraham, Isaac and Jacob), nor the accounts of the conquest of the land by Joshua after the Exodus of the Children of Israel from Egypt and the wanderings in the desert have any archaeological or other evidence to back them up, and researchers assume that these events never occurred. The *we* forms here are all inclusive and relate to the two researchers involved in the interaction, and they shape the relations between them – relations of agreement and lack thereof in the textual context of the debate as well as beyond it. Thompson for his part supplies expressions of agreement for and approval of Dever's remarks.

However, immediately afterwards, the debate heats up and spills over into personal recriminations. At this point, it becomes clear to the reader, thanks to a comment by the editor, that Thompson had worked under Dever at the excavations in the Gezer site in 1967. As I described at the beginning of Section 5.6 above, this site is mentioned in the Bible as being one of the cities built by King Solomon. Thompson makes the accusation that Dever's excavations at Gezer were based a priori on the biblical text, i.e. he accuses him of what Finkelstein calls 'Bible archaeology' (see Section 5.5 above). Dever, however, denies this charge and hurls a number of accusations of his own at Thompson:

(5) Dever: […] on the internet, *you* or somebody else is accusing *me* of going to Gezer to find the Solomonic gate. Now that is absolutely ridiculous. *Our* arguments have always been on straight archaeological lines: ceramic evidence, stratigraphy, historical points that can be fixed […]. To say that *we* went to Gezer looking for the Solomonic gate is really slanderous. Maybe *you* didn't write it.

Thompson: *I* did write something similar to that. *I* talked about what *we* went to Gezer for.

Dever: *You* knew that *we* were not looking for the –

Thompson: *I* knew that *we* were looking for Solomon's gate.

Dever: Well, that's something *I* didn't know. That's what *I* mean by slander. Do not impute motives to *other people* when *you* don't know what *they're* doing.

Thompson: *I'm* talking about what *you* told *me* and what *we* did together. And that's what *I've* written.

Dever: Dear sir, if *you* have not read what *I've* written about Biblical archaeology for these last 30 years…

Thompson: *I* have.

Dever: *You* know very well that *I* have been the most outspoken foe of that kind of simple-minded Biblicism. *I've* been the most outspoken, and *I* have paid for it dearly. To say that *our* archaeological strategy comes out of the Bible is really nonsense. Tom, *I* don't care in the least whether Solomon ever existed. *I'm* probably more of a disbeliever than *you*. *I* don't really care about the tradition. *I* don't believe any of the myths. But as a historian, an archaeologist, *I* believe the date of the construction of the six-chambered gate [at Gezer] is a historical and archaeological matter. If the Bible had never been written, *I* still would need to date that gate.

Thompson: *I* was reconstructing a piece of the history of scholarship. *You*, who were very, very central to that piece of the history of scholarship, deny what *I* say.

Dever: *You* were reconstructing this from *your* memory. *I* only ask *you* to go back and read what *I* was writing in the '60s.

Thompson: That's quite fair.

Dever: Don't hold *me* responsible for a memory *you* have of what *you* thought *I* was doing 30 years ago.   (Panel: 33)

Dever responds here to what he perceives as an attack on his ethos by a counter-attack on that of Thompson, especially where scientific integrity is concerned: that he is imprecise in what he writes and then refuses to stand behind what he wrote, reconstructs inaccurately from memory and is not sufficiently conversant with the research literature. The expression "really nonsense" is a very harsh phrase. I found nothing that resembles it in the written scientific literature.

The first-person singular pronouns and second-person pronouns in this case reflect the direct confrontation between the two researchers and underscore the position that each one holds. In contrast, numerous first-person plural pronouns go beyond the boundaries of the interaction. Thompson uses this form to denote the work he and Dever shared in the past. However, Dever's use of the first-person plural pronoun is not inclusive: He uses it to relate to groups of researchers with whom he worked, but does not necessarily include Thompson in it, for example in the sentence: "*Our* arguments have always been on straight archaeological lines."

Dever also uses the general phrase 'other people' in order to accuse Thompson of improper behavior. Dever's direct attacks on Thompson continue afterwards too, and the previous accusations are joined by additional ones that target Thompson's credibility, his research capability and his very standing as a researcher:

(6)   Dever: Furthermore, *your* argument about the population of Judah is
                        incorrect. [...] it's simply wrong to say that Judah doesn't have a
                        significant population before 701.
        Thompson: Oh, *I* estimate about 2,000.
        Dever: In a recent article *you* said, I'm quoting you, "There were only a
                        few dozen people in Judah." *I* will bring you the quotation.
        Thompson: Please do.
        Dever: *I* shall. *I* don't misquote. And if *you* look at any of the studies of
                        demography that the Israelis have done, then *you* know that this
                        is quite wrong.
        Thompson: The figure of 2,200 is cited from Finkelstein's most recent article.
        Dever: *You* said a few dozen. Many of *your* facts are wrong, Tom,
                        because *you* do not control the archaeological data. No Israeli
                        archaeologist has bothered to answer *you* – Tom, *I'm* sorry to say
                        this – because none takes *you* seriously. Not a single one. They
                        have not answered *you* because *you* get your facts wrong, Tom.
                        You need to check with archaeologists.
        Lemche: That's nonsense. *We* have to get in here because of an attack –
        Dever: When *a man* denies *he's* written what *he's* written *I* do not trust
                        him any longer.   (Panel: 35)

Dever provides here a series of pejoratives aimed at Thomson: His arguments are
"incorrect" or "simply wrong," he does not "control the archaeological data" and
cannot be trusted. These are extremely harsh and offensive comments, and so it
comes as no surprise that Lemche, who shared Thompson's research and like him
is a follower of the "Copenhagen school," feels it necessary to intervene and defend
him. Lemche's use of *we* is not entirely clear; perhaps he includes the editor and
discussion moderator, who in accordance with the conventions of an academic
debate could be expected to step in to prevent an *ad hominem* attack of this kind.

    Dever however uses neutral phrases here such as 'a man', similar to 'other
people' in the previous example, in order to give his argument a more general and
decisive nature: According to this, Thompson's behavior, which he has described,
is intolerable, not only on a personal level but also on the generally ethical one,
based on the accepted norms as they are practiced among the members of the
scientific community.

    Thompson, for his part, uses a more sophisticated weapon, that of ironic im-
plication. In the context of the archaeological findings in Jerusalem, he says:

(7)   Dever: You can't say there's nothing in the tenth century. You can only
                        say in her [= Margreet Steiner's] publication there's not.
        Thompson: I'm not talking about the metaphysical existence of the reality of
                        the past. I'm talking about the basis of evidence.   (Panel: 36)

The use of negative structure of "I'm not talking" implies that there is someone who is talking about it, hinting at Dever. In this way, the phrase "the metaphysical existence of the reality of the past" is perceived as echoing the belief that seemingly underlies what Dever is saying. Whereas Dever is trying to persuade his interlocutors that he sticks only to the archaeological facts, Thompson is indirectly accusing him of not distinguishing between concrete reality and metaphysics, a serious accusation in a scientific context.

In another case, there is an implied accusation through the ambiguous use of the first-person plural pronoun:

(8)   I said *we* have no knowledge of any Jerusalem. *We* work only with evidence,
      Bill.   (Thompson, Panel: 35)

The pronoun 'we' in the first sentence can be interpreted as being aimed at the entire research community; however, in the second sentence it is ambiguous. On the one hand, it refers to researchers who belong to Thompson's school of thought, that they are the ones who "work only with evidence." On the other, one can see a modal intent and interpret it as targeting all researchers that seek to reach the truth. Based on the accepted standards of research, everyone needs to "work only with evidence," and Thompson is insinuating that not everyone does, an implied accusation of Dever, as if to say that his claims are not based on evidence.

In an oral discussion of this nature, it is not an accepted technique to cite from written texts. However, another strategy is being demonstrated here, one that is less accepted in papers in journals (although it can be found in them too), which is to cite private conversations with other researchers in order to bolster one's claims. Here is an example:

(9)   Benjamin Sass, who may be the greatest expert on this, says they're fakes. He
      even told me who made them, of course at a dinner party, never in print.
                                                                  (Lemche, Panel: 38)

The person who is being quoted is given a positive evaluation in order to bolster his status as an authority. In another place, on the question of the possibility of identifying forgeries, Lemche enlists a conversation held with a researcher whom he presents as one whose credibility is beyond question to support him. In another place, he quotes another private conversation, this time regarding the findings in Jerusalem:

(10) Last year at a conference in Jerusalem on exactly this period, David Ussishkin
      said to me, "It's not only an argument from silence. There's not a single sherd
      [piece of pottery] from the tenth century." Ussishkin is a very conservative

scholar. I know him quite well. He really wants to retain the Davidic monarchy. Not a single sherd, not a single one belongs to the tenth century.

(Lemche, Panel: 35)

Since the revisionists in this debate are the target of criticism because they do not
base themselves sufficiently on archeology, Lemche uses a quotation from personal
conversation with the Israeli archaeologist David Ussishkin of Tel Aviv University.
To bolster his own argument, Lemche adds comments regarding Ussishkin's positions, which he ostensibly knows from first-hand experience. Because Ussishkin's
views, according to Lemche, are conservative rather than revisionist, his remarks
seem to contradict his own ideology. This claim is intended to bolster the validity
of his own comments and add further credence to their objective nature.

In response to these remarks by Lemche, Dever responds in the same coin
and notes his own personal acquaintance with Ussishkin and other Israeli
archaeologists.

(11)   When Ussishkin says not a single sherd belongs to the tenth century, that's
because he dates all of it to the ninth century. I grew up with this generation
of archaeologists. They're my best friends. I know these guys, and I know
what they think. And I can tell you that not a single one of the other Israeli
archaeologists agrees with this Low Chronology, except Israel Finkelstein.

(Dever, Panel: 35–36)

Dever tries to weaken Lemche's position by claiming to have a deeper acquaintance and of many more years with the Israeli archaeologists, who disagree among
themselves over the dating of the findings (as shown in Section 5.6 above). It
appears that for the participants in this debate, it is important to have a personal
acquaintance and hold private conversations with Israeli archaeologists who are
located in physical proximity to the relevant sites, do the work "in the field" and
see the findings with their own eyes. The personal acquaintance serves to increase
the credibility of the researcher who is located further away and reinforces his
authority.

The importance of eye-witness testimony comes in Example (2) above too. In
it, Dever notes the fact that he held the finding in his hands and saw it with his
own eyes as a way to bolster his claim that the inscription that was found at Tel
Dan and was inscribed with the words *Beit David* is not counterfeit. He continues
to defend the authenticity of the inscription:

(12)   I was there shortly after it was found. I've known Biran [director of the excavation at Tel Dan] for 40 years. The woman who found it, Gila Cook, I hired
at Hebrew Union College. I have handled the inscription. I know what I'm
talking about. There's no way.   (Dever, Panel: 37)

Here too, Dever shores up his claim not only by means of eye-witness testimony, but also by pointing to his person acquaintance of many years with the relevant researchers, which ostensibly makes him an authority on their credibility. Another example of bolstering the claim by means of pointing to the personal circumstances of the speaker can be found in the following remarks by Thompson:

> (13) When I talked about there not being anything in Jerusalem, I wasn't referring to Finkelstein and I wasn't referring to Ussishkin. I was talking about Margaret Steiner's publications of the Kenyon excavations. She's publishing with the Copenhagen international seminar, of which I'm the editor.
> (Thompson, Panel: 36)

The fact that Thompson is the editor of Steiner's publications can imply that he is more familiar with their content than others. It seems, however, that the purpose of his mentioning his status as being responsible for a major publication here is mainly to support his own ethos as having standing in the research community, and not necessarily to directly support his claims.

It is notable that all the examples just cited mention other researchers – but without any citation from them. Consequently, one can say that with the exception of a few places, the dialogue of the speakers with the research community is fairly limited here. The face-to-face presence and nature of the spoken (as opposed to written) interaction causes them to relate mainly to those who are present in the room, at the expense of the more accepted strategy in written scientific communication.

The research community, its conventions and interests are mentioned mainly by means of first-person plural pronouns. First, each of the participants does so in order to refer to his own specific research community. Dever does so to refer to archaeologists:

> (14) *We* archaeologists are not trying to prove these early stories to be historical. *We* get accused of it, but we're not doing it. (Dever, Panel: 32)

Lemche uses this form to refer to the revisionist Bible researchers, in response to the moderator's request to explain the position of the group:

> (15) *We* have no problem with Israel as such, with the name. *We* have problems with the interpretation of the name. (Lemche, Panel: 40)

In Thomson's remarks too we can see the use of the pronoun *we* to refer to the group as a whole, although he hesitates to ascribe position that he expresses to the entire group and prefers to be circumspect when making generalizations:

(16)   *We* do deny the existence, *at least I do*, of a united monarchy in the tenth century – for a number of reasons.   (Thompson, Panel: 34)

McCarter represents yet a third discipline in this debate – the moderator of the debate introduced him as an expert paleographer, and it is in this capacity that he turns to him and asks him to express his opinion on the authenticity of the inscriptions from Tel Dan and Ekron. The first-person plural pronouns in his remarks below can therefore be interpreted as referring to paleographers, those whose expertise involves deciphering inscriptions:

(17)   I tend to doubt the Ekron inscription is a forgery. When *we* have a chance to study it, questions may arise. It's too new for *us* to have the confidence in it that *we* have in the Tel Dan inscription.   (McCarter, Panel: 38)

These uses of the plural first-person pronouns have a twofold function: On the one hand, they unite and construct solidarity with a research group or specific discipline. At the same time, they separate, because the drawing of the boundaries of the group distinguishes it from the rest, and especially the members of the group from researchers from other disciplines.

In another place, Lemche uses the form *we* in order to reference the national or cultural collective to which he belongs, the Danes:

(18)   It's a kind of history you find in North tradition, among the Vikings, my forefathers. *We* love those stories, but *we* don't believe them to be true.
                                                                (Lemche, Panel: 32)

The use Lemche makes of *we* in this example essentially separates: His aim is to distinguish between the Danes, who in his opinion relate to their mythological ancestors in the proper fashion (they love them but do not believe they actually existed), and the Israelis or Jews, who tend to believe in the historical existence of their mythological forbearers, even when in his view, there is nothing to justify this in the historical research.

However, many of the *we* forms are inclusive and the speakers use them to appeal to all those who aspire to reach the scientific truth, who have the knowledge and findings, available to all those interested in explaining and drawing conclusions from it. This knowledge can include archaeological findings, as proposed in this example:

(19)   For example *we* have the seal impressions of Berakhyahu Ben Neriyahu […].
                                                                (Lemche, Panel: 37)

But it can also be scientific knowledge of another kind that can be taken into consideration, as McCarter proposes in this example:

(20)   *We* do have some evidence. *We* have the evidence of tradition.

(McCarter, Panel: 36)

The second-person pronoun *you* may be generic and refer to a broader group outside the specific interaction. The use of the indefinite *you* is informal (Quirk et al. 1972: 222), and is typical of face-to-face interaction. This is the case in Example (18) above, as well as in the following example:

(21)   We don't rely just on the gates; it's a whole complex. It's pottery, it's tomb styles, it's house styles, it's gates, it's fortifications. When *you* put it together I think *you* can make a very good case for a tenth-century state.   (Dever, Panel: 42)

The pronoun *you* as it is used here refers to anyone observing the data who can draw conclusions from them. The invitation to the addressee to participate here has a rhetorical aim, which is to make him a partner to the speaker's conclusion.

Face-to-face interaction is a dialogic situation that is essentially different from the written scientific communication held on the pages of journal articles. Although its characteristics are not the focus of interest in this study, I have presented an example here in order to examine it on the background of the academic papers that we looked at in the previous sections. To sum up this section, I will note a number of points that come up when comparing written and spoken communication, without getting into the numerous obvious differences between them.

The issue of citations is an interesting one. The researchers participating in the symposium that we looked at are authors of academic papers by virtue of their academic status, and consequently, it is clear that they have full mastery of the practice of how to cite from the relevant scientific literature, an element of major importance in the writing of an academic paper. Nevertheless, in the oral interaction on the very same subjects that they deal with in their studies, they made reference to very few academic papers. In fact, only one example of this could be found, in Example (6) above. In the current situation, they mostly cited private conversations, and most of the claims that were raised were not attributed to a particular author. The comparison between this interaction and the academic article clearly demonstrates the difference that Bakhtin noted between primary and secondary genres. The oral discussion here is the primary, simple genre, with the scientific paper being the secondary genre, into which other previous texts have been embedded. The dialogicity, which is a feature shared by all the texts, takes on a different form in each of the genres within the same discourse community.

Another important issue is that of politeness. Hunston (2005) described the 'conflict articles' as those that address unique conventions of politeness, which in her view are expressed mainly in the fact that the criticism expressed in them is aimed more directly at the researchers themselves, whereas ordinary articles tend to direct their criticism towards positions or findings, so as to minimize the harm caused to the face of researchers holding the opposing view. This difference was intensified in the face-to-face interaction, which was defined in advance as being adversarial. This kind of interaction is absent the distance that is typical of written communication (Chafe 1985, see Section 2 above), and consequently, it contained utterances that were blunt and direct to an extent we do not see in written scientific communication, not even in conflict articles.

An interesting difference was found in the way the first-person plural pronoun was used. The way the debate was planned and held, as well as in response to the moderator's instructions, each of the speakers identified himself as representing a particular discipline or specific group of researchers: archaeologists, paleographers, Bible scholars of the minimalist school. The adversarial situation invited considerable use of the exclusive *we*, which did not include the other participants in the interaction, creating an effect of distinguishing each group from the others. These pronouns served to demonstrate the self-identity of the speakers, draw the boundaries of the group to which they belong and distinguish between them and the other speakers.

CHAPTER 6

Conclusions

The academic paper is a masterpiece of sophisticated use of previous texts and of integration of various voices together with that of the author, of appealing to different audiences and anticipating their reactions. The dialogicity is evident in the academic paper in various forms and to different degrees of explicitness, while the components that shape it are essentially the same. The dialogicity model proposed in Section 4.6 served me in Chapter 5 to analyze entire papers including all their dialogic elements. The various components that I presented in the model appear in all the papers and were shown to be significant in shaping the dialogic nature of the discourse.

Citations from the research literature hold a dialogue with the discourse community and specific members of it. They present the new study as being part of the entire scientific enterprise and help to underscore the innovation that it contains. They help the author to occupy a place within the disciplinary community by defining his self-identity as a researcher and to reinforce his authority and credibility as such. Integral citations underscore the place occupied by the cited researchers, and thus also the dialogue that the author is holding with them. Consequently, they have the potential to increase the confrontational nature of the paper. And indeed, we have seen that the preference for non-integral citation may be used to diminish the confrontational tone of the paper, a fact that became evident in the analysis of the paper written in the classic pattern (Section 5.3 above).

All the papers of the confrontational pattern analyzed in Chapter 5 contained concession structures. This is the quintessential means used to allow the voice of the other to be heard in the text and to confront it. On the other hand, the concession structure may hold a participatory dialogue with the reader too, who may entertain certain objections to the arguments as he reads, as well as with the discourse community. Sometimes knowledge shared by the members of the community is presented in the satellite, and its mention in the text serves as an endorsement of the shared assumptions and conventions, thereby contributing to disciplinary solidarity.

The first-person plural pronouns may be exclusive or inclusive. When they are inclusive (i.e. include the addressee), they hold a participatory dialogue with the reader and the discourse community. When they are exclusive (i.e. do not include the addressee), they foreground the positions of the authors and their ethos.

When a single speaker uses them, he may do so to create a distinction between the group that he belongs to and other participants in the discourse, thereby distancing them and intensifying the conflict between them and himself.

Questions create a dialogue with the reader, share the thinking process with him and push him in the direction the author is going. Questions that appear at the end of an article and that aim at future research hold a dialogue with the disciplinary community in that they are aimed at its shared project. On the other hand, rhetorical questions may contribute to the intensification of the adversarial nature of the discourse: They serve to refute another's claims and even to present them as unreasonable, especially when they take on an added ironic note.

All these devices, as noted, were found in the analysis of the complete papers that I presented in Chapter 5, and their role in creating dialogicity has been shown to be central and significant. In addition, as expected, occurrences of further elements from Hyland's model (2005) were found, as well as of Fløttum (2006), as I showed in Section 4.1, for example, the obligation modals that reflect dialogicity with the reader. However, these were few and far between and have not been systematically analyzed in this study. At the same time, I included them in the table below, which is aimed at presented picture of all the linguistic devices that shape scientific dialogicity that is as complete as possible.

In the classic pattern (Section 5.3), a number of linguistic devices that serve to minimize the confrontation were found: the use of non-integral references and clusters of references and the mention of opposing views using of non-specific expressions. However, we also saw that the presence of the author in the text is diminished by means of the use of non-integral references. First-person plural pronouns are inclusive and create a feeling of participation with the reader and the discourse community.

However, when analyzing the papers of the confrontational pattern, it was evident how all these tools were mobilized to confront opponents. Extensive quotations from opponents are cited in order to expose the flaws in their arguments, and when the confrontation intensifies, the citations from other sources decrease. Concession structures intensify the contrast between the positions of the different sides. Exclusive pronouns underscore the positions of the authors in contrast to those they seek to refute.

When the confrontation becomes personal, the author can use an alternative strategy: Instead of turning to the particular researcher whose work he is impugning, he can turn to the entire disciplinary community and ask it to objectively decide the dispute between them.

The confrontational patterns revealed additional tools to express criticism: First, just as Hunston predicted, there is a more extensive use of a value system. Expressions that belong to such a system, and which represent researchers from both

Table 7. Scientific dialogicity: A detailed description

Dialogue with				
	Reader		Concession (reference to an assumed claim or reservation) Inclusive *we* for reader's participation Non-genuine questions Indefinite pronouns, 'the reader' Imperatives ('note', 'consider', 'let-us') Obligation modals 'It is (adjective) to do' Metatext	Manifest intertextuality
	Discourse community	Specific member	Citations Concession (reference to a specific claim) Negative evaluation Punctuation for expressing reservation (quotation marks, exclamation marks) Irony (including ironic rhetorical questions) Labeling *Ad hominem* arguments	
		Non-specific member	Non-attributed opinions ('in most books') Concession (reference to a non-attributed specific claim)	
		The community as a whole	Concession (reference to shared knowledge and norms) Communal *we* Open questions directed to the scientific project Obligation modals 'It is (adjective) to do' Positive/negative evaluation (referred to shared norms and values)	
			Use of genre conventions	Constitutive intertextuality

sides and their arguments as either "good" or "bad," on the one hand, hold a direct dialogue with the opposing researchers, while on the other, hold an indirect dialogue with the discourse community via its values systems and accepted norms.

A further rhetorical tool is ironic implication, which can also be created by means of certain punctuation marks, such as quotation marks and exclamation points, to express reservations.

My model for scientific dialogicity proposes that a distinction be made first and foremost between an appeal to the reader and an appeal to the discourse community. While in the case of an academic paper, the potential reader will almost always be a member of the discourse community, the appeal to him appears to take on a special nature, one that is worthy of distinction. Inviting the reader to participate in the reading and drawing of conclusions is an important rhetorical tool, which has a unique effect in helping the author to achieve his goals.

As noted, the integrated model suggests that a distinction be made between a particular member of the discourse community and a non-specific member. I consider this an important distinction as well. The appeal to a specific member of the community in a written text is in fact carried out in only one way: by referring directly to his work; whereas the appeal to the discourse community employs varied textual means and is far more subtle in nature.

Another possibility, one that we find in confrontational articles, is that the specific member of the community is a researcher at whom the entire text is aimed – at his work, his claims, ways of thinking, character, etc. In this case, the conventional nature of the scientific text is altered, and the model may be expanded to include a selection of further linguistic devices aimed to directly attack the rival researcher, as we see in Table 7 above. These are rhetorical devices that rhetoric has always recognized, and they underscore the nature of the article as a persuasive text.

Bibliography

Adams Smith, Diana. 1987. "Variation in field related genres." *ELR Journal* 1: 10–32.

Ahmad, Ummul K. 1997. "Research article introductions in Malay: Rhetoric in an emerging research community." In *Culture and Styles in Academic Discourse,* Anna Duszak (ed.), 273–303. Berlin: Mouton de Gruyter.

Allen, Graham. 2000. *Intertextuality.* London and New York: Routledge.

Amossy, Ruth. 2002. "Introduction to the study of doxa." *Poetics Today* 23 (3): 369–394.

Anscombre, Jean-Claude. 1985. "Grammaire traditionelle et grammaire argumentative de la concession." *Revue Internationale de Philosophie* 39 (4): 333–349.

Anscombre, Jean-Claude. (ed.). 1995. *Théorie des topoi.* Paris: Kimé.

Aristotle. 1982. *Rhetoric.* John H. Freese (trans.), Cambridge: MA.

Authier-Revuz, Jacqueline. 1982. "Hétérogénéité montrée et hétérogénéité constitutive: éléments pour une approche de l'autre dans le discours." *DRLAV Revue de Linguistique* 26: 91–151.

Azar, Moshe. 1997. "Concession relations as argumentation." *Text* 17 (3): 301–316.

Azar, Moshe. 1999. "On concession in contemporary Hebrew." In *Studies in Ancient and Modern Hebrew in Honour of M. Z. Kaddari,* Shimon Sharvit (ed.), 285–304. Bar Ilan University Press [In Hebrew].

Bakhtin, Mikhail M. 1981. *The Dialogic Imagination: Four Essays*, Emerson, Caryl and Holquist, Michael (trans.), Michael Holquist (ed.). Austin, Texas: University of Texas Press.

Bakhtin, Mikhail M. 1986. "The problem of speech genres." In *Speech Genres and Other Late Essays,* Vern W. McGee (trans.). Austin. Texas: University of Texas Press.

Bartholomae, David. 1986. "Inventing the university." *Journal of Basic Writing* 5: 4–23.

Baynham, Mike. 1999. "Double-voicing and the scholarly 'I': On incorporating the words of others in academic discourse." *Text* 19 (4): 485–504.

Bazerman, Charles. 1984. "Modern evolution of the experimental report in physics: Spectroscopic articles in *Physical Review,* 1893–1980." *Social Studies in Science* 14: 163–196.

Bazerman, Charles. 1988. *Shaping Written Knowledge: The Genre and Activity of the Experimental Article in Science.* Madison: University of Wisconsin Press.

Becher, Tony. 1981. "Towards a definition of a disciplinary culture." *Studies in Higher Education* 6 (2): 109–122.

Becher, Tony and Trowler, Paul R. 2001. *Academic Tribes and Territories: Intellectual Enquiry and the Cultures of Disciplines* (2nd edition). Buckingham: Open University Press/SRHE.

Beller, Mara. 1999. *Quantum Dialogue: The making of a Revolution.* Chicago/London: University of Chicago Press.

Berge, Kjell Lars. 2003. "The scientific text genres as social actions: Text theoretical reflections on the relations between context and text in scientific writing." In *Academic Discourse: Multidisciplinary Approaches,* Kjersti Fløttum and François Rastier (eds), 141–157. Oslo: Novus.

Berkenkotter, Carol and Huckin, Thomas. 1995. *Genre Knowledge in Disciplinary Communication: Cognition/Culture/Power.* Hillsdale: Lawrence Erlbaum.

Biber, Douglas, Johansson, Stig, Leech, Geoffrey and Conrad, Susan. 1999. *Grammar of Spoken and Written English.* London: Longman.

Bloch, Joel and Chi, Lan. 1995. "A comparison of the use of citations in Chinese and English academic discourse." In *Academic Writing in a Second Language: Essays on Research and Pedagogy,* Diane D. Belcher and George Braine (eds), 135–155. Norwood, NJ: Ablex Publishing Corporation.

Bloor, Thomas. 1996. "Three hypothetical strategies in philosophical writing." In *Academic Writing: Intercultural and Textual Issues,* Eija Ventola and Anna Mauranen (eds), 19–43. Amsterdam: John Benjamins.

Bondi, Marina. 2002. "Attitude and episteme in academic discourse: Adverbials of stance across genres and moves." *Textus* XV: 249–264.

Bondi, Marina. 2004. "'If you think this sounds very complicated, you are correct': Awareness of cultural differences in specialized discourse." In *Intercultural Aspects of Specialized Communication,* Christopher N. Candlin and Maurizio Gotti (eds), 53–77. Bern: Peter Lang.

Bondi, Marina. 2005. "Metadiscursive practices in academic discourse: Variations across Genres and Disciplines." In *Dialogue within Discourse Communities,* Julia Bamford and Marina Bondi (eds), 3–30. Tubingen: Niemeyer.

Bondi Marina, Gavioli, Laura and Silver, Marc. 2004. "Introduction". In *Academic Discourse, Genre and Small Corpora,* Marina Bondi, Laura Gavioli and Marc Silver (eds), 7–13. Rome: Officina edizioni.

Breeze, Ruth. 2010. "They say, we do. Writers' strategic positioning in the discourses of political communication research." In *Constructing Interpersonality: Multiple Perspectives on Written Academic Genres,* Rosa Lorés-Sanz, Pilar Mur-Dueñas and Enrique Lafuente-Millán (eds), 163–180. Cambridge Scholars Publishing.

Brett, Paul. 1994. "A genre analysis of the Results section of sociology articles." *English for Specific Purposes* 13 (1): 47–59.

Brown, Penelope and Levinson, Stephen C. 1987. *Politeness: Some Universals in Language Usage.* Cambridge: Cambridge University Press.

Burgess, Sally. 2002. "Packed houses and intimate gathering: Audience and rhetorical structure." In *Academic Discourse,* John Flowerdew (ed.), 196–215. Harlow: Longman.

Chafe, Wallace. 1982. "Integration and involvement in speaking, writing, and oral literature." In *Spoken and Written Language,* Deborah Tannen (ed.). Norwood, NJ: Ablex.

Chafe, Wallace. 1985. "Linguistic Differences produced by differences between speaking and writing." In *Literacy, Language and Learning,* David R. Olson, Nancy Torrance and Angela Hyldyard (eds). Cambridge: Cambridge University Press.

Cherry, Roger D. 1988. "Ethos vs. persona: Self representation in written discourse." *Written Communication* 5: 251–276.

Chittleborough, Philip and Newman, Morgan E. 1993. "Defending the term 'argument'." *Informal Logic* 15 (3): 189–207.

Clark, Burton R. 1962. *Faculty Culture.* Berkeley: Center for the Study of Higher Education, University of California.

Čmejrková, Svetla. 1996. "Academic writing in Czech and English." In *Academic Writing: Intercultural and Textual Issues,* Eija Ventola and Anna Mauranen (eds), 137–153. Amsterdam: John Benjamins.

Čmejrková, Svetla. 2007. "Intercultural dialogue and academic discourse." In *Dialogue and culture*, Marion Grein and Edda Weigand (eds), 73–94. Amsterdam: John Benjamins.

Čmejrková, Svetla and Daneš, Františ. 1997. "Academic writing and cultural identity: The case of Czech academic writing." In *Culture and Styles in Academic Discourse*, Anna Duszak (ed.), 41–62. Berlin: Mouton de Gruyter.

Compton, Arthur Holly. 1923. "A quantum theory of the scattering of X-rays by light elements." *Physical Review* 21: 483–502.

Coulmas, Florian. (ed.). 1986. *Direct and indirect speech*. Berlin/New York: Mouton.

Crevels, Mily. 2000. *Concession: A Typological Study. Doctoral Dissertation*, Amsterdam: University of Amsterdam.

D'Angelo, Frank J. 1976. "The search for intelligible structure in the teaching of composition." *College Composition and Communication* 27: 142–147.

Dascal, Marcelo and Katriel, Tamar. 1977. "Between semantics and pragmatics: The two types of 'but' – Hebrew 'Aval' and 'Ela'." *Theoretical Linguistics* 4: 143–172.

Dressen, Dacia F. 2002. "Identifying textual silence in scientific research articles: Recontextualization of the field account in Geology." *Hermes, Journal of Linguistics* 28: 81–107.

Dudley-Evans, Tony. 1989. "An outline of the value of genre analysis in LSP work." In *Special Language: From Humans Thinking to Thinking Machines*, Christer Laurén and Marianne Nordman (eds), 72–79. Clarendon, Philadelphia: Multilingual Matters.

Duszak, Anna. 1994. "Academic discourse and intellectual styles." *Journal of Pragmatics* 21: 291–313.

Duszak, Anna. 1997. "Cross-cultural academic communication: A discourse-community view." In *Culture and Styles in Academic Discourse*, Anna Duszak (ed.), 11–40. Berlin: Mouton de Gruyter.

Fairclough, Norman. 1992. *Discourse and Social Change*. Cambridge: Polity Press.

Fleck, Ludwik. 1979. *Genesis and Development of a Scientific Fact*, Thaddeus J. Trenn and Robert K. Merton (trans.). Chicago and London: University of Chicago Press.

Fløttum, Kjersti. 2003. "Bibliographical references and polyphony in research articles." In *Academic Discourse: Multidisciplinary Approaches*, Kjersti Fløttum and François Rastier (eds), 97–119. Oslo: Novus.

Fløttum, Kjersti. 2003. "Bibliographical references and polyphony in research articles." In *Academic Discourse: Multidisciplinary Approaches*, Kjersti Fløttum and François Rastier (eds), 97–119. Oslo: Novus.

Fløttum, Kjersti. 2004. "Traces of others in research articles: The citation cluster." In *New Directions in LSP studies*, Khurshid Ahmad and Margaret Rogers (eds), 153–159. Guilford: University of Surrey.

Fløttum, Kjersti, Dahl, Trine and Kinn, Torodd. 2006. *Academic Voices*. Amsterdam/Philadelphia: John Benjamins.

Fortanet-Gómez, Inmaculada. 2004. "The use of 'we' in university lectures: Reference and function." *English for Specific Purposes* 23: 45–66.

Fredrickson, Kirsten M. and Swales, John M. 1994. "Competition and discourse community: Introductions from *Nysvenska studier*." In *Text and Talk in Professional Contexts*, Britt-Louise Gunnarsson, Per Linell and Bengt Nordberg (eds), 9–21. Uppsala: ASLA (The Swedish Association of Applied Linguistics).

Frumuşelu, Mihai D. 2007. "Linguistic and argumentative typologies of concession: An integrative approach." In *Proceedings of the Sixth Conference of the International Society for the Study of Argumentation*, Frans H. Van Eemeren, Anthony J. Blair, Charles A. Willard and Bart Garssen (eds), 425–431. Amsterdam, International Center for the Study of Argumentation.

Geertz, Clifford. 1988. *Works and Lives: The Anthropologist as Author*. Stanford: Stanford University Press.

Genette, Gérard. 1997. *Palimpsests: Literature in the Second Degree*, Channa Newman and Claude Doubinsky (trans.). Lincoln/London: University of Nebraska Press.

Giannoni, Davide S. 2005. "Negative evaluation in academic discourse: A comparison of English and Italian research articles." *Linguistica e Filologia* 20: 71–99.

Goffman, Erving. 1981. *Forms of Talk*. Oxford: Blackwell.

Goldenberg, Gideon. 1998. "On verbal structure and the Hebrew verb." In *Studies in Semitic Linguistics: Selected Writings*, 148–196. Jerusalem: Magnes Press.

Groom, Nick. 2000. "Attribution and averral revisited: Three perspectives on manifest intertextuality in academic writing." In *Patterns and Perspectives: Insights into EAP Writing Practices*, Paul Thompson (ed.). Reading: Reading University Press.

Groom, Nick. 2005. "Pattern and meaning across genres and disciplines: An exploratory study." *Journal of English for Academic Purposes* 4(3): 257–277.

Gross, Alan G. 1985. "The form of the experimental peper: A realisation of the myth of induction." *Journal of Technical Writing and Communication* 15 (1): 15–26.

Halliday, Michael A. K. 2004. *The Language of Science*. London/New York: Continuum.

Hamel, Rainer E. 2007. "The dominance of English in the international scientific periodical literature and the future of language use of science." In *Linguistic inequality in scientific communication today, AILA Review 20*, Augusto Carli and Ulrich Ammon (eds), 53–71. Modena-Reggio Emilia/Duisburg-Essen.

Harwood, Nigel. 2005. "We do not seem to have a theory… The theory I present here attempts to fill this gap: Inclusive and exclusive pronouns in academic writing." *Applied Linguistics* 26 (3): 343–375.

Harwood, Nigel. 2007. "Political scientists on the functions of personal pronouns in their writing: An interview-based study of 'I' and 'We'." *Text & Talk* 27 (1): 27–54.

Holmes, Richard. 1997. "Genre analysis and the social sciences: An investigation of the structure of research article discussion sections in three disciplines". *English for Specific Purposes* 16: 321–337.

Huckin, Thomas N. 1993. "Surprise value in scientific discourse", Paper presented at the 9th European Symposium on Language for Special Purposes, Bergen, 2–6 August 1993. Prepublication draft.

Huckin, Thomas. 2002. "Textual silence and the discourse of homelessness." *Discourse & Society* 13 (2): 347–372.

Hunston, Susan. 1993a. "Evaluation and ideology in scientific writing." In *Register analysis: Theory and practice*, Mohsen Ghadessy (ed.). London: Pinter.

Hunston, Susan. 1993b. "Professional conflict: Disagreement in academic discourse." In *Text and Technology: In Honour of John Sinclair*, Mona Baker, Gill Francis and Elena Tognini Bonelli (eds), 115–134. Amsterdam: John Benjamins.

Hunston, Susan. 1994. "Evaluation and organization in a sample of written academic discourse." In *Advanced in Written Text Analysis*, Malcolm Coulthard (ed.). London: Routledge.

Hunston, Susan. 2004. "'It has rightly been pointed out...': Attribution, consensus and conflict in academic discourse." In *Academic Discourse, Genre and Small Corpora*, Marina Bondi, Laura Gavioli and Marc Silver (eds), 15–33. Rome: Officina edizioni.

Hunston, Susan. 2005. "Conflict and consensus: Construing opposition in Applied Linguistics." In *Strategies in Academic Discourse*, Elena Tognini Bonelli and Gabriella Del Lungo Camiciotti (eds), 1–15. Amsterdam/Philadelphia: John Benjamins.

Hyland, Ken. 1998a. *Hedging in scientific research articles*. Amsterdam/Philadelphia: John Benjamins.

Hyland, Ken. 1998b. "Persuasion and context: The pragmatics of academic metadiscourse." *Journal of Pragmatics* 30: 437–455.

Hyland, Ken. 1999. "Academic attribution: Citation and the construction of disciplinary knowledge." *Applied Linguistics* 20 (3): 341–365.

Hyland, Ken. 2000. *Disciplinary Discourses: Social Interaction in Academic Writing*. London: Longman.

Hyland, Ken. 2001a. "Bringing in the reader: Addressee features in academic articles." *Written Communication* 18: 549–574.

Hyland, Ken. 2001b. "Humble servants of the discipline? Self-mention in research articles." *English for Specific Purposes* 20: 207–226.

Hyland, Ken. 2002a. "Authority and invisibility: Authorial identity in academic writing." *Journal of Pragmatics* 34: 1091–1112.

Hyland, Ken. 2002b. "What do they mean? Questions in academic writing." *Text* 22 (4): 529–557.

Hyland, Ken. 2003. "Self-citation and self-reference: Credibility and promotion in academic publication." *Journal of American Society for Information Science and Technology* 54 (3): 251–259.

Hyland, Ken. 2005. "Stance and engagement: A model of interaction in academic discourse." *Discourse Studies* 7: 173–192.

Hyland, Ken. 2006. "Disciplinary differences: Language variations in academic discourses." In *Academic discourse across disciplines*, Ken Hyland and Marina Bondi (eds), 17–45. Bern: Peter Lang.

Hyland, Ken. 2009. "Constraints vs. creativity: Identity and disciplinarity in academic writing." In *Commonality and Individuality in Academic Discourse*, Maurizio Gotti (ed.), 25–52. Bern: Peter Lang.

Illi, Cornelia. 1994. *What Else Can I tell You? A Pragmatic Study of English Rhetorical Questions as Discursive and Argumentative Acts*. Stockholm: Almqvist & Wiksell International.

Ivanič, Roz. 1998. *Writing and identity: The discoursal construction of identity in academic writing*. Amsterdam: John Benjamins.

Jacoby, Sally. 1987. "Reference to other researchers in literary research articles." *ELR Journal* 1: 33–78.

Jørgensen, Charlotte. 2009. "Interpreting Perelman's *Universal Audience*: Gross versus Crosswhite." *Argumentation* 23: 11–19.

Knorr-Cetina, Karin D. 1981. *The Manufacture of Knowledge: Toward a Constructivist and Contextual Theory of Science*. Oxford: Pergamon.

Kourilová, Magda. 1996. "Interactive function of language in peer reviews of medical papers written by NN users of English." *UNESCO-ALSED LSP* 19: 4–21.

Kuo, Chih Hua. 1999. "The use of personal pronouns: Role relationships in scientific journal articles." *English for Specific Purposes* 18 (2): 121–138.

Kuhn, Thomas. 1962. *The Structure of Scientific revolutions*. Chicago: University of Chicago Press.

Kristeva, Julia. 1980. *Desire in Language: A Semiotic Approach to Literature and Art*, Thomas Gora, Alice Jardine and Léon S. Roudiez (trans.), Léon S. Roudiez (ed.). New York: Columbia University Press.

Labov, William. 1972. "The transformation of experience in narrative syntax." In *Language in the Inner City*, 354–396. Philadelphia, PA: University of Pennsylvania Press.

Lafuente-Millán, Enrique, Mur-Dueñas, Pilar, Lorés-Sanz Rosa and Vázquez-Orta, Ignacio. 2010. "Interpersonality in written academic discourse: Three analytical perspectives." In *Constructing Interpersonality: Multiple Perspectives on Written Academic Genres*, Rosa Lorés-Sanz, Pilar Mur-Dueñas and Enrique Lafuente-Millán (eds), 13–40. Cambridge Scholars Publishing.

Latour, Bruno. 1987. *Science in Action: How to Follow Scientists and Engineers through Society*. Milton Keynes: Open University Press.

Latour, Bruno and Woolgar, Steve. 1979. *Laboratory Life: The Social Construction of Scientific Facts*, Beverly Hills: Sage.

Lewin, Beverly A., Fein, Jonathan and Young, Lynne. 2001. *Expository Discourse: A Genre-based Approach to Social Science research texts,* London and New York: Continuum.

Lindeberg, Ann-Charlotte. 2004. *Promotion and Politeness: Conflicting Scholarly Rhetoric in Three Disciplines*. Åbo Akademis Förlag: Åbo Akademi University Press.

Linell, Per. 1998. "Discourse across boundaries: On the recontextualization and the blending of voices in professional discourse". *Text* 18 (2): 143–157.

Livnat, Zohar. 2005. "La rhétorique de l'objectivité. le rôle de l'auteur dans l'écriture scientifique." *Questions de Communication* 9: 95–121.

Livnat, Zohar. 2010a. "Impersonality and grammatical metaphors in scientific discourse: The rhetorical perspective." *LIDIL* 41: 103–119.

Livnat, Zohar. 2010b. *The Rhetoric of Scientific Papers: Language and the Discourse Community*. Bar-Ilan University Press. [In Hebrew]

Loffler-Laurian, Anne-Marie. 1980. "L'expression du locuteur dans les discours scientifiques. 'JE', 'NOUS' et 'ON' dans quelques textes de chimie et de physique." *Revue de Linguistique romane* 44: 135–157.

Maingueneau, Dominique. 1987. *Nouvelles tendances en analyses du discours*. Paris: Hachette.

Mann, William and Thompson, Sandra A. 1986. "Relational propositions in discourse." *Discourse Processes* 9: 57–90.

Mann, William and Thompson, Sandra A. 1988. "Rhetorical Structure Theory: Towards a functional theory of text organization." *Text* 8 (3): 243–281.

Martín-Martín, Pedro. 2005. *The Rhetoric of the Abstract in English and Spanish Scientific Discourse*. Bern: Peter Lang.

Mauranen, Anna. 1993. "Contrastive ESP rhetoric: metatext in Finnish-English economics texts." *English for Specific Purposes* 12: 3–22.

Montgomery, Scott L. 1996. *The Scientific Voice*. New York/London: Guilford Press.

Motta-Roth, Désirée. 1998. "Discourse analysis and academic book reviews: A study of text and disciplinary cultures." In *Studies in English for Academic Purposes* (Vol. 9), Immaculada Fortanet, Santiago Posteguilloo, Juan Carlos Palmer and Juan Francisco Coll (eds), 29–59. Filología. Universitat Jaume I: Collecció Summa.

Mühlhäusler, Peter and Harré, Rom. 1990. *Pronouns and People: The Linguistic Construction of Social and Personal Identity*. Oxford: Blackwell.

Myers, Greg. 1985. "Texts as knowledge claims: The social construction of two biology articles." *Social Studies of Science* 15: 593–630.

Myers, Greg. 1989. "The pragmatics of politeness in scientific articles." *Applied Linguistics* 10 (1): 1–35.

Myers, Greg. 1992. "'In this paper we report...': Speech acts and scientific facts." *Journal of Pragmatics* 17: 295–313.

Nuyts, Jan. 2001. *Epistemic Modality, Language, and Conceptualization*. Amsterdam/Philadelphia: John Benjamins.

Nystrand, Martin. 1986. *The structure of Written Communication: Studies in Reciprocity between Writers and Readers*. Orlando: Academic Press.

Pera, Marcello. 1994. *The Discourses of Science*, Clarissa Botsford (trans.). Chicago/London: The University of Chicago Press.

Pennycook, Alastair. 1994. "The politics of pronouns." *ELT Journal* 48 (2).

Perelman, Chaïm. 1982. *The Realm of Rhetoric*, William Kluback (trans.). Notre Dame: University of Notre Dame Press.

Perelman, Chaïm and Olbrechts-Tyteca, Lucie. 1958. *The New Rhetoric: A Treatise on Argumentation*, John Wilkinson and Purcell Weaver (trans.). Notre Dame: University of Notre Dame Press, 1969.

Piazza, Roberta. 2002. "The pragmatics of conductive questions in academic discourse." *Journal of Pragmatics* 34: 509–527.

Poppi, Franca. 2004. "Writer representation and authorial stance: A case study of the introductory chapters in economics textbooks." In *Academic Discourse, Genre and Small Corpora*, Marina Bondi, Laura Gavioli and Marc Silver (eds), 127–137. Rome: Officina edizioni.

Poudat, Celine and Loiseau, Sylvain. 2005. "Authorial presence in academic genres." In *Strategies in Academic Discourse*, Elena Tognini Bonelli and Gabriella Del Lungo Camiciotti (eds), 51–68. Amsterdam/Philadelphia: John Benjamins.

Prelli, Laurence J. 1989. *A Rhetoric of Science: Inventing Scientific Discourse*. University of South California Press.

Quirk, Randolph, Greenbaum, Sidney, Leech, Geoffrey and Svartvik, Jan. 1972. *A Grammar of Contemporary English*. London: Longman.

Resinger, Hildegard. 2010. "Same science, same stance? A cross-cultural view on academic positioning in research articles." In *Constructing Interpersonality: Multiple Perspectives on Written Academic Genres*, Rosa Lorés-Sanz, Pilar Mur-Dueñas, and Enrique Lafuente-Millán (eds), 205–218. Cambridge Scholars Publishing.

Roubrieux, Jean-Jacque. 1993. *Elements de Rhetorique d'Argumentation*. Paris: Dunod.

Rounds, Patricia L. 1987. "Multifunctional personal pronoun use in an educational setting." *English for Specific Purposes* 6 (1): 13–29.

Salager-Meyer, Françoise. 1998. "The rationale behind academic conflict: From outright criticism to contextual 'niche' creation." *UNESCO-ALSED LSP* 21: 4–23.

Salager-Meyer, Françoise. 1999. "Contentiousness in written medical English discourse: A diachronic study (1810–1995)." *Text* 19: 371–398.

Salager-Meyer, Françoise. 2001. "The bittersweet rhetoric of controversiality in 19th and 20th-century French and English medical literature." *Journal of Historical Pragmatics* 2: 141–173.

Salager-Meyer, Françoise, Alcaraz-Ariza, María Ángeles and Zambrano, Nahirana. 2003. "The scimitar, the dagger and the glove: Intercultural differences in the rhetoric of criticism in Spanish, French and English medical discourse (1930–1995)." *English for Specific Purposes* 22 (3): 223–247.

Samson, Christina. 2004. "Some interpersonal meta-discourse aspects in contemporary written economics lectures." In *Academic Discourse, Genre and Small Corpora*, Marina Bondi, Laura Gavioli and Marc Silver (eds), 71–85. Rome: Officina Edizioni.

Sasaki, Tsuguya. 2007. "The place of Modern Hebrew as a lingua franca of Jewish studies." *Language Problems & Language Planning* 31 (2): 131–141.

Searle, John R. 1969. *Speech Acts: An Essay in the Philosophy of Language*. Cambridge: Cambridge University Press.

Shapin, Steven. 1984. "Pump and circumstance: Robert Boyle's literary technology." *Social Studies in Science* 14: 481–520.

Silver, Marc and Bondi, Marina. 2004. "Weaving voices: A study of article openings in historical discourse." In *Academic Discourse – New Insights into Evaluation*, Gabriella Del Lungo Camiciotti and Elena Tognini Bonelli (eds), 141–159. Bern: Peter Lang.

Sinclair, John M. 1988. "Mirror for a text." *Journal of English and Foreign Languages* 1: 15–44.

Sperber, Dan and Wilson, Deirdre. 1981. "Irony and the use-mention distinction." In *Radical Pragmatics*, Peter Cole (ed.), 295–318. New York: American Press.

Stotesbury, Hilkka. 2006. "Gaps and false conclusions: Criticism in research article abstracts across the disciplines." In *Academic discourse across disciplines*, Ken Hyland and Marina Bondi (eds), 123–148. Bern: Peter Lang.

Swales, John M. 1986. "Citation analysis and discourse analysis." *Applied Linguistics* 7 (1). 39–56.

Swales, John M. 1990. *Genre Analysis: English in Academic and Research Settings*. Cambridge/New York/Melbourne: Cambridge University Press.

Swales, John M., Ahmad, Ummul K., Chang, Yu Ying, Chaves, Daniel, Dressen, Dacia F. and Seymour, Ruth. 1998. "Consider this: The role of imperatives in scholarly writing." *Applied Linguistics* 19 (1): 97–121.

Sweetser, Eve. 1990. *From Etymology to Pragmatics*. Cambridge: Cambridge University Press.

Tang, Romana and Suganthi, John. 1999. "The 'I' in identity: Exploring writer identity in student academic writing through the first person pronoun." *English for Specific Purposes* 18 (supplement 1): S23–S39.

Taylor, Charles. 1991. "The dialogic self." In *The interpretive turn: Philosophy, Science, Culture*, David R. Heily and James F. Bohman (eds), 304–314. Cornell University Press.

Taylor, Gordon and Chen, Tingguang. 1991. "Linguistic, cultural and subcultural issues in contrastive discourse analysis: Anglo-American and Chinese scientific texts." *Applied Linguistics* 12 (3): 319–336.

Thompson, Geoff. 1996. "Voices in the text: Discourse perspectives on language reports." *Applied Linguistics* 17 (4): 501–530.

Thompson, Geoff. 2001. "Interaction in academic writing: Learning to argue with the reader." *Applied Linguistics* 22 (1): 58–78.

Thompson, Geoff and Hunston, Susan. 2001. "Evaluation: An introduction." In *Evaluation in Text: Authorial Stance and the Construction of Discourse*, Susan Hunston and Geoff Thompson (eds), 1–27. Oxford: Oxford University Press.

Thompson, Geoff and Thetela, Puleng. 1995. "The sound of one hand clapping: The management of interaction in written discourse." *Text* 15 (1): 103–127.

Thompson, Paul. 2005. "Aspects of identification and position in intertextual reference in PhD Theses." In *Strategies in Academic Discourse*, Elena Tognini Bonelli and Gabriella Del Lungo Camiciotti (eds), 31–50. Amsterdam/Philadelphia: John Benjamins.

Thompson, Susan. 1997. "Why ask questions in monologue? Language choice at work in scientific and linguistic talk." In *Language at Work: Selected papers from the Annual Meeting of the British Association for Applied Linguistics*, Susan Hunston (ed.). University of Birmingham: Clarendon, Multilingual Matters.

Toulmin, Stephen. 1958. *The uses of Argument*. Cambridge: Cambridge University Press.

Tutin, Agnès. 2010. "Evaluative adjectives in academic writing in the humanities and social sciences." In *Constructing Interpersonality: Multiple Perspectives on Written Academic Genres*, Rosa Lorés-Sanz, Pilar Mur-Dueñas and Enrique Lafuente-Millán (eds), 219–242. Cambridge Scholars Publishing.

Van Eemeren, Frans H. and Grootendorst, Rob. 1995. "Perelman and the fallacies." *Philosophy and Rhetoric* 28: 122–133.

Van Eemeren, Frans H., Grootendorst, Rob and Snoeck Henkemans, Francisca. 2002. *Argumentation Analysis, Evaluation, Presentation*. Mahwah, NJ: Lawrence Erlbaum.

Vold, Eva Thue. 2006. "The choice and use of epistemic modality markers in linguistics and medical research articles." In *Academic discourse across disciplines*, Ken Hyland and Marina Bondi (eds), 225–249. Bern: Peter Lang.

Vološinov, Valentin N. 1973. "Reported speech." In *Reading in Russian Poetics: Formalist and Structuralist Views*, Ladislav Matejka and Krystyna Pomorska (eds), 149–175. Ann Arbor, MI: Michigan Slavic Publication.

Wales, Katie. 1980. "Exophora re-examined: The uses of the personal pronoun 'WE' in present-day English." *UEA Papers in Linguistics* 12: 21–44.

Wales, Katie. 1996. *Personal Pronouns in Present-Day English*. Cambridge: Cambridge University Press.

Webber, Pauline. 1994. "The function of questions in different medical English genres." *English for Specific Purposes* 13: 257–268.

Weddle, Perry. 1978. *Argument: A Guide to Critical Thinking*. New-York: McGraw-Hill Book Company.

Weigand, Edda. 2009. *Language as Dialogue*. Amsterdam/Philadelphia: John Benjamins.

Weizman, Elda. 1984. "Some register characteristics of discourse structure in journalistic language." *Applied Linguistics* 5 (1): 39–50.

Weizman, Elda. 2011. "Conveying indirect reservation through discursive redundancy." *Language Sciences* 33 (2): 295–304.

Wierzbicka, Anna. 1974. "The semantics of direct and indirect discourse." *Papers in Linguistics* 7 (3–4): 267–307.

Appendix
Corpus of journal articles

A. Social Sciences (for Chapters 3 and 4)

These papers were published in Hebrew. Here, I will give their names in Hebrew, immediately followed by a translation of the title as it appears in the English title-page of the publication.

1. Yariv Tsfati, "*hashed aadati beisrael: betokh habakbuk – al esh ktana*" (= Isreal's ethnic demon: Inside the bottle, on a slow flame), *Megamot* 40 (1) 1999: 5–27.
2. Samuel Shye and Gil Goldzweig, "*yetsiratiut keharchava shel inteligentsia: hagdarat shatchot vehasharot mivniot*" (= Creativity as an extension of intelligence: Faceted definition and structural hypotheses), *Megamot* 40 (1) 1999: 31–53.
3. Tamar Kennet-Cohen, Shmuel Bronner and Carmel Oren, "*nituach-al shel tokef hanibuy shel markivey maarekhet hamiun launiversitaot beisrael klapey midat haatslacha balimudim*" (= A meta-analysis of the predictive validity of the selection process used by universities in Israel), *Megamot* 40 (1) 1999: 54–71.
4. Samia Dawud-Noursi, Michael Lamb and Kathleen Sternberg, "*hashpaat alimut gufanit bamishpacha al histaglut yeladim beveit hasefer*" (= The effects of domestic violence on children's adjustment at school), *Megamot* 40 (1) 1999: 72–102.
5. Nissan Rubin and Drora Peer, "*tiksey prisha mitsahal – tkasim rishmiim utkasim pratiim*" (= Army retirement rites: Formal and informal), *Megamot* 40 (1) 1999: 103–130.
6. Gila Menachem and Idit Gejst, "*safa, taasuka vezika leisrael bekerev oley CIS bishnot ha-90*" (= Hebrew language proficiency, occupation and attachment to Israel among immigrants from the former Soviet Onion in the 1990s), *Megamot* 40 (1) 1999: 131–148.
7. Ayelet Kohn, "*hair veʿhair*': *yitsug hametsiut hachevratit bemekomon*" (= 'The city': The presentation of social reality in a local weekly), *Megamot* 40 (1) 1999: 149–166.
8. Ilana Brosch and Yochanan Peres, "*kamut mul ʿeikhutʾ shel yeladim – dilemma klalit bemusagim isreeliim*" (= Child quantity versus 'quality': A general dilemma in Israeli terms), *Megamot* 40 (2) 2000: 185–192.
9. Uzi Ben-Shalom and Gabriel Horenczyk, "*zehut tarbutit vehistaglut bekerev nearim olim beproyekt naale 16*" (= Cultural identity and adaptation among participants in the Naale 16 project), *Megamot* 40 (2) 2000: 199–217.
10. Isaac A. Friedman, "*lechatsey hatafkid baavodat menahel beit hasefer kemenbey shchika*" (= Role pressures in school principal's work as predictors of burnout), *Megamot* 40 (2) 2000: 218–243.
11. Vered Tzur and Amalya Oliver-Lumerman, "*morim veyitsugam: modaot chipus ovdim kemeshakfot havnaia profesionalit*" (= Presenting teachers: Professional construction as reflected in employment ads), *Megamot* 40 (2) 2000: 244–261.

12. Dalia Rachman-Moore and Nira Danziger, "*hevdeley migdar bitchilat hakariera hamiktsoit shel bogrey minhal asakim*" (= Gender differences in early career attainment of business school graduates, *Megamot* 40 (2) 2000: 262–279.

13. Victor Florian, Asa Kasher and Ruth Malkinson, "*hityachasut haprat, hatsibur vehatik-shoret lemishpachot shakulot beisrael: seker data kahal*" (= Public and Media perception of bereaved families in Israel: A national survey), 280–297.

14. Edna Granit and Liron Nathan, "*kehilot virtualiot: mivne chevrati chadash? *" (= Virtual communities: A new social structure?), *Megamot* 40 (2) 2000: 298–315.

15. Gideon M. Kressel, "*shvitat hastudentim umodel hatriada shel Georg Simmel*" (= The student's strike and Simmel's model of triadic relationship), *Megamot* 40 (2) 2000: 316–322.

16. Rivka Bar-Yosef, "*yaldut basviva hachevratit hamishtana shel sof haelef hasheni*" (= Childhood in a changing society at the end of the second millennium), *Megamot* 40 (3) 2000: 365–381.

17. Malca Aleck and Yoel Yinon, "*hashpaat sug hazehut shel hayachid vehakvutsa al midat tokfanutam*" (= Aggressive behavior as a function of identity type among individuals and groups), *Megamot* 40 (3) 2000: 382–412.

18. Yisraela Herer and Ofra Mayseless, "*histaglut rigshit vechevratit etsel mitbagrim baaley dfus shel hipuch tafkidim bamishpacha*" (= Emotional and social adjustment of adolescents who show role-reversal in the family), *Megamot* 40 (3) 2000: 413–441.

19. Liron Dushnik and Naama Sabar Ben-Yehoshua, "*bein kavana leyekholet: dilemot etiot shel morim beisrael shel shnot ha-90*" (= Intention vs. abilities – the ethical dilemmas facing Israeli teachers in the 90s), *Megamot* 40 (3) 2000: 442–465.

20. Gabriel Weimann, "*migdar upirsomet. nashim ugvarim betashdirey hapirsomet hatelevizionit beisrael*" (= Gender differences in Israeli TV commercials), *Megamot* 40 (3) 2000: 466–485.

21. Zehava Rosenblatt and Ayala Ruvio, "*i-bitachon taasukati bekerev morim bachinukh haal-yesodi beisrael: gisha rav-memadit*" (= Job insecurity of Israeli secondary school teachers: A multidimensional approach), 486–511.

22. Yair Amichai Hamburger and Shaul Fox, "*irgunim virtualiim iskiim bainternet: mahut ir-gunit chadasha*" (= Virtual business organizations on the internet: A new organizational model), *Megamot* 40 (3) 2000: 512–530.

23. Gariel Cavaglion, "*hayeled kehavnaya tarbutit: hadugma shel hachinukh hamini vehaasbara haminit bayishuv hayehudi bitchilat hamea ha-20*" (= Childhood as a social construction: The case of sex education in the Jewish settlement of the early 20th century), *Megamot* 40 (3) 2000: 531–548.

24. Edna Lomsky-Feder and Tamar Rapaport, "*lehishaer baarets o laazov? – itgur haetos hat-sioni besipurey hagira*" (= Immigrants challenge the national ethos: Jewish-Russian students deconstruct Zionism), *Megamot* 40 (4) 2000: 571–590.

25. Yechezkel Dar and Shaul Kimhi, "*tfisa atsmit shel bigur beikvot sherut hachova betsahal*" (= Military service and self-perceived maturation among Israeli youth), *Megamot* 40 (4) 2000: 591–616.

26. Nava Maslovaty, "*darkhey hitmodedut im dilemot chevratiot umusariot beveit hasefer haye-sodi hamamlakhti-dati: haadafot hamorim veifyuney hamorim*" (= Teaching strategies for coping with socio-moral dilemmas in religious-state elementary schools: Teachers' preferences and teachers' characteristics), *Megamot* 40 (4) 2000: 617–635.

27. Audrey Addi-Raccah, "*hashpaat meafyeney hamorim umeafyeney beit hasefer al hekef mis-rat haoraa shel gvarim venashim*" (= The effect of individual and school characteristics on gender differences in teachers' working hours), *Megamot* 40 (4) 2000: 636–659.

28. Bracha Kramarski, "*meat-kognitsia ufituach hayekholet liftor beayot matematiot hamut-sagot besituatsia muchashit uvesituatsia mufshetet*" (= Metacognition and the ability to solve math problems presented in concrete and abstract contexts), *Megamot* 40 (4) 2000: 660–685.

29. Meir Freshtman, "*drakhim lezihuy chavurot (cohesive groups) bereshet chevratit*" (= Cohesive groups detection in a social network), *Megamot* 40 (4) 2000: 686–705.

30. Moshe Semyonov, Noah Lewin-Epstein and Hadas Mandel, "*madad meudkan lestatus sotsio-economi shel mishlchey-yad beisrael*" (= Updated socioeconomic scale for occupations in Israel), *Megamot* 40 (4) 2000: 706–718.

B. Archaeology (for Chapter 5)

1.   Margreet Steiner, "The archaeology of ancient Jerusalem", *Currents in Research: Biblical Studies* 6, 1998, pp. 143–168.

2.   Nadav Na'aman, "The contribution of the Amarna letters to the debate on Jerusalem's Political Position in the Tenth Century B.C.E.", *Bulletin of the American Schools of Oriental Research* 304, 1996, pp. 17–27.

3.   Eilat Mazar, "*sridey armon david hamelech birushalayim: mechkar bearcheologia mikrait*" (= Remains of King David's palace in Jerusalem: Research in Biblical archaeology), *New studies on Jerusalem* 2, 1996, pp. 9–15.

4.   Eilat Mazar, "King David's palace", *Biblical Archaeology Review* 23, 1997, pp. 52–57, 74.

5.   Eilat Mazar, "*haomnam gilinu et armon hamelech david?*" (= Did we find King David's palace?), *New Studies on Jerusalem* 11, 2006, pp. 7–16.

6.   Eilat Mazar, "Did I find King David's palace?" *Biblical Archaeology Review* 32, 2006, pp. 17–27, 70.

7.   Israel Finkelstein, Ze'ev Herzog, Lily Singer-Avitz and David Ussishkin, "Has King David's palace in Jerusalem been found?", *Tel-Aviv* 34 (2), 2007, pp. 142–164.

8.   Israel Finkelstein, "The archaeology of the United Monarchy: an Alternative View", *Levant* 28, 1996, pp. 177–187.

9.   Amihai Mazar, "Iron Age Chronology: A Reply to I. Finkelstein", *Levant* 29, 1997, pp. 157–167.

10. Israel Finkelstein, "Bible Archaeology or Archaeology of Palestine in the Iron Age? A Rejoinder", *Levant* 30, 1998, pp. 167–174.

11. "Biblical minimalists meet their challengers", *Biblical Archaeology Review* 23, 1997, pp. 27–42, 66.

Author index

A

Adams Smith, Diana 51, 114, 117
Ahmad, Ummul 4, 21, 124
Allen, Graham 10–13
Amossy, Ruth 38
Anscombre, Jean-Claude 38, 68
Aristotle 93
Authier-Revuz, Jacqueline 13
Azar, Moshe 68–72

B

Bakhtin, Mikhail 9–13, 16, 17, 18, 47, 49, 52, 59, 103, 110, 120, 192
Bartholomae, David 93
Baynham, Mike 50, 59
Bazerman, Charles 23–24, 25, 50, 51, 52–53
Becher, Tony 21, 51, 52, 93
Beller, Mara 1, 23, 25, 44, 64
Berge, Kjell 22
Berkenkotter, Carol 29, 45
Biber, Douglas 1, 51, 110, 135, 140
Bloch, Joel 4, 124
Bloor, Thomas 51
Bondi, Marina 21, 28, 50
Breeze, Ruth 51
Brett, Paul 28
Brown, Penelope 125
Burgess, Sally 4, 124

C

Chang, Yu Ying 21
Chafe, Wallace 7, 110, 193
Chaves, Daniel 21
Chen, Tingguang 4, 124
Cherry, Roger 93
Chi, Lan 4, 124
Chittleborough, Philip 69–70

Clark, Burton 21
Čmejrková, Svetla 4, 51, 124
Conrad, Susan 1, 51, 110, 135, 140
Coulmas, Florian 16
Crevels, Mily 68

D

Dahl, Trine 4, 21, 22, 48–49, 50, 51, 52, 54, 57, 60, 88, 90, 94–95, 102–103, 106, 134, 135, 137, 147
Dascal, Marcelo 67
Dressen, Dacia 21, 25
Dudley-Evans, Tony 28, 45
Duszak, Anna 4, 124

F

Fairclough, Norman 13
Fein, Jonathan 28, 42, 85
Fleck, Ludwik 1
Fløttum, Kjersti 4, 21, 22, 48–49, 50, 51, 52, 54, 57, 60, 62, 64, 88, 90, 94–95, 102–103, 106, 134, 135, 137, 147, 196
Fortanet-Gómez, Inmaculada 51, 101
Fredrickson, Kirsten 4
Frumuşelu, Mihai 66–67, 68, 71–72

G

Gavioli, Laura 28
Geertz, Clifford 159–160
Genette, Gérard 13–14
Giannoni, Davide 124
Goffman, Erving 17, 59–60
Goldenberg, Gideon 96
Greenbaum, Sidney 135, 192
Groom, Nick 21, 51
Grootendorst, Rob 18, 44
Gross, Alan 26

H

Halliday, Michael 8, 40, 42
Hamel, Rainer 3
Harré, Rom 101
Harwood, Nigel 21, 51, 94
Holmes, Richard 21
Huckin, Thomas 23, 25, 29, 45
Hunston, Susan 27, 28, 50, 65, 81, 125–127, 140, 160, 165, 192, 196
Hyland, Ken 21, 22, 23, 24, 26, 28, 29, 42, 48–49, 50, 51, 53, 55, 61, 62, 93, 95, 100, 102, 104, 106, 110, 111–112, 114–115, 116, 119, 123, 134, 136, 142, 196

I

Illi, Cornelia 111, 113
Ivanič, Roz 51

J

Jacoby, Sally 50
Johansson, Stig 1, 51, 110, 135, 140
Jørgensen, Charlotte 18, 19

K

Katriel, Tamar 67
Kinn, Torodd 4, 21, 22, 48–49, 50, 51, 52, 54, 57, 60, 88, 90, 94–95, 102–103, 106, 134, 135, 137, 147
Knorr-Cetina, Karin 23, 25
Kourilová, Magda 123
Kuo, Chih Hua 21, 51, 101, 106
Kuhn, Thomas 21, 29
Kristeva, Julia 12–13, 17

L

Labov, William 99
Lafuente-Millán, Enrique 4

Latour, Bruno   23, 25, 33, 36, 37, 39, 40, 42
Leech, Geoffrey   1, 51, 110, 135, 140, 192
Levinson, Stephen   125
Lewin, Beverly   28, 42, 85
Lindeberg, Ann-Charlotte   21, 23, 28, 45, 119
Linell, Per   25
Livnat, Zohar   9, 32, 40, 85, 95, 99, 119
Loffler-Laurian, Anne-Marie   101
Loiseau, Sylvain   51
Lorés-Sanz, Rosa   4

M

Maingueneau, Dominique   13
Mann, William   68, 70
Martín-Martín, Pedro   4, 51, 101, 109, 124–125, 134, 135, 136, 141, 159
Mauranen, Anna   4
Montgomery, Scott   25, 50, 52, 113, 118
Motta-Roth, Désirée   123
Mühlhäusler, Peter   101
Mur-Dueñas, Pilar   4
Myers, Greg   21, 24, 39, 47–48, 51, 53, 94, 95, 102, 103, 106, 125, 134–135, 148–149

N

Newman, Morgan   69–70
Nuyts, Jan   42
Nystrand, Martin   7

O

Olbrechts-Tyteca, Lucie   18–19, 22, 34–35, 39, 40, 41, 81, 102, 180–181

P

Pera, Marcello   22, 36
Pennycook, Alastair   102
Perelman, Chaïm   18–19, 22, 34–35, 39, 40, 41, 43, 81, 102, 180–181
Piazza, Roberta   111
Poppi, Franca   51
Poudat, Celine   51
Prelli, Laurence   22

Q

Quirk, Randolph   135, 192

R

Resinger, Hildegard   4
Roubrieux, Jean-Jacque   91
Rounds, Patricia   101

S

Salager-Meyer, Françoise   4, 124
Samson, Christina   21
Sasaki, Tsuguya   3
Searle, John   111
Seymour, Ruth   21
Shapin, Steven   23
Silver, Marc   28, 50
Sinclair, John   17
Snoeck Henkemans, Francisca   44
Sperber, Dan   17
Stotesbury, Hilkka   21, 124
Suganthi, John   51
Svartvik, Jan   135, 192
Swales John   4, 21, 22, 25, 26, 27–28, 45, 50, 51, 53–54, 55, 73, 114, 116, 118, 124
Sweetser, Eve   7–8

T

Tang, Romana   51
Taylor, Charle   7
Taylor, Gordon   4, 124
Thetela, Puleng   7, 117
Thompson, Geoff   7, 16–18, 49, 59, 89, 107, 117, 121, 126
Thompson, Paul   50
Thompson, Sandra   68, 70
Thompson, Susan   51
Toulmin, Stephen   44
Trowler, Paul   21
Tutin, Agnès   4

V

Van Eemeren, Frans   18, 44
Vázquez-Orta, Ignacio   4
Vold, Eva Thue   21
Vološinov, Valentin   59

W

Wales, Katie   101
Webber, Pauline   51, 111, 114, 117
Weddle, Perry   173
Weigand, Edda   15, 18
Weizman, Elda   62
Wierzbicka, Anna   16
Wilson, Deirdre   17
Woolgar, Steve   23, 25, 33, 36, 37, 39, 40, 42

Y

Young, Lynne   28, 42, 85

Subject index

A

addressivity   1, 11
adversatives   49, 50, 66–67, 73
adversativity   67
allusion   14
animator   59–60
architextuality   14
argumentative direction   67–68
argumentum ad hominem   22,
   178, 197
argumentum e silentio   140, 142
attitude markers   48–49
authorship   39–40, 59–63
averral   17

B

black box   37
boosters   48–49

C

cognitive actions   9, 95–96
coherence   33, 65–66, 168–169
communal *we*   106–110, 120–121
concession
   argumentative   70–71
   circumstantial   70–71
   content   68, 71–72
   direct-rejection   69, 76
   epistemic   68
   indirect rejection   69–70
   speech act   68, 72
   textual   68
conflict article   126–127, 160–
   161, 192–193
conventionalized silence   25
conventions   13, 18, 125–126, 133
   disciplinary   19, 47–48, 55,
      94, 120, 121, 195, 197
   stylistic   13, 20, 49
credibility (of the researcher)
   93, 159–160, 189, 195
   see also ethos

D

disciplinary community   21–22
   membership in   19, 21, 49,
      78, 93, 99–100, 106, 109,
      118, 119
disciplinary variations   21,
   44, 53
doxa   38

E

emotion   95, 114, 178
enthymeme   38
ethos   66, 93–94, 99–100, 150,
   160, 174, 181, 195
   see also credibility (of the
      researcher), self-identity
evaluation   75, 95, 114, 119, 124,
   126–127, 196–197
   see also value system

G

genre   4, 11, 20, 21–22, 52
   primary   47, 192
   secondary   12, 47, 192
   variations   8, 11, 13, 59, 70,
      110, 123–124, 192
grammatical metaphor   40

H

hedges   24, 42, 43, 48–49, 85,
   99, 124
   see also modals
hypertextuality   14

I

imperatives   49, 197
impersonality   9, 40–41, 94, 99,
   119, 124, 140, 159
IMRD scheme   9, 25–27, 28
indefinite pronoun   48–49, 135,
   139–140, 192, 197

innovation   28, 29, 33–34, 64,
   65, 86–87, 125, 195
intertextuality
   constitutive   13, 48, 50, 51,
      120, 121, 197
   horizontal dimension of
      12–13, 47
   manifest   13, 16, 47, 48, 50,
      51, 52, 121, 197
   vertical dimension of   12–13,
      47, 50
irony   154, 167, 173, 177, 181, 187,
   196, 197

M

meta-text   48–49, 197
metatextuality   14
modals   24, 42, 119
   see also hedges
   obligation   49, 136–137, 140,
      144, 156, 176, 196, 197
monologue-dialogue continuum
   1, 7

N

negation   49, 134, 147
neutral observer   180–181
niche   27, 28, 30–31, 64, 73–76,
   114, 118, 119, 124
   see also research space,
      territory
nominalization   8–9

O

objective reader   180, 196
objectivity   40, 95, 125, 133,
   139–140, 149, 159, 160, 172–
   173, 180–182

P

paratext   14
passive   9, 40, 95, 134–135,
   139–140

plagiarism   14
politeness   53, 125, 134–135,
   148–149, 192–193
polyphony   25, 48
   explicit   52, 88
   implicit   88
   internal   89
principal   59–60
punctuation   155, 197
   see also quotation marks

Q
question
   non-genuine   113, 114–117,
      119–120, 121, 197
   open   117–118, 119, 121, 197
   real   111, 113–114, 117–118,
      119–120
   rhetorical   111–113, 146, 147,
      196, 197
quotation marks   62, 148,
   149–150, 153–154, 197

R
rationality   18, 43–44
reported speech   16–17
research space   27, 64–65,
   73–76, 93, 124, 132
responsive understanding   11–
   12, 47, 103
rhetorical structure theory   68,
   70

S
self-identity   66, 193, 195
self-mention   48–49
shared knowledge   18, 20,
   37–39, 44, 49, 50, 109, 120,
   121, 142, 143–144, 197
solidarity   49, 50, 79, 92, 107,
   109–110, 195
speech-act
   domain   8
   concession   68, 72
   initiative   15–16
   reactive   15–16
   representative   15–16, 72

T
territory   27, 29, 45, 73, 118
   see also niche, research space
topos   38, 68–72, 76
translation   63
transtextuality   13

U
universal audience   18–19,
   34–35, 43–44
unspecifiable other   17, 107
unspecified other   17, 89

V
value
   negative   81, 174, 175, 178
   positive   144, 149
   system   126–127, 196–197